The American Dream/ The American Nightmare:

The Authoritative Guide to Building Your Custom Home

John L. Cogdill, Jr.
and
J. J. Cogdill

The American Dream/
The American Nightmare:

The Authoritative Guide to Building Your Custom Home

By
John L. Cogdill, Jr.
And
J.J. Cogdill

Published by:
Keystone Publishing Company
108 Industrial Loop North
Orange Park, FL 32073 U.S.A.

Authors:
John L. Cogdill, Jr.
J.J. Cogdill

Illustrator:
Jennifer Cogdill

Cover Design:
Empire Communications Group

Editors:
Jennifer Cogdill
Jerry Cogdill
Kelly Bowman
Kim Cole

All rights reserved. No part of this book may be reproduced or transmitted in any form or by any means, electronic or mechanical, including photocopying, recording, or by any information storage and retrieval system without the express written permission of the author, except for the inclusion of brief quotations in a review.

Copyright © 2000 by John L. Cogdill, Jr. and J.J. Cogdill, All Rights Reserved

First Printing 2000

Printed in the United States of America
Printing by: Continental Printing

Library of Congress Number has been requested.

ISBN #: 0-9678684-3-2

<u>Acknowledgments</u>

The authors of this book wish to express their sincere gratitude and thanks to the following people, without whom this work would not have been possible:

- **Jerry Cogdill**: loving wife and mother whose hard work and support helped us through the toughest of times and inspired us always, whether we wanted to be inspired or not.

- **Jennifer Cogdill**: talented daughter and sister whose artistic flights of fancy helped make this book more enjoyable for all of us.

- **Dr. Leon Wetherington**: one of the best professors you could ever have. He is indeed a credit to our profession and we are honored to have him as one of our own.

- **Kelly Bowman**: one of the most gracious and beautiful women the world has ever known. Her love and compassion is felt by everyone around her and I hope her spark for life will be with us always.

- **Laura McGrady**: mother-in-law and grandmother who showed love and class during the best and worst of times. A true inspiration if ever the word applied to a person, she helped make all that we ever do possible.

- **Randy Wilder, PE**: thank you for all of your help with our land developing projects and we hope to see much more of you in the years to come.

- **Jim Lucas, PE**: thank you for all of your help in our land developing adventures and we wish you continued success in the future.

- **The faculty and staff of the M.E. Rinker Sr. School of Building Construction at the University of Florida**: without you, this work would literally have not been possible. Thank you for giving us the tools to grow and thrive in the industry of our choice and please keep up the good work. The construction industry needs all the skilled professionals you can give us.

- **Les Loggins, John Bracey**: and all the gang at Empire Communications. Thank you for helping us endure the trials and tribulations of producing this book and we hope you can continue to help us endure the hardships of the harsh world of advertising in the future.

- **Renay Daigle**: thank you for all your help. Your boundless enthusiasm has seen us through many trials, and I hope with the birth of your daughter that your enjoyment of life reaches new heights.

- **Jim & Shirley Duduit**: and the folks at Continental Printing. We hope we have not set a world record for setting and breaking printing deadlines, although we may be pretty close. Thank you for putting up with us and being not only one of our avid literary fans, but also one of our best clients.

- **Kim Cole**: the most unique sister-in-law anyone could have. Your skills in the mortgage business are legendary, as is your temperance and self control.

<u>Warning/Disclaimer</u>

This book has been written for the purpose of providing information on designing and building a custom home. The authors and the publisher of this book are in no way providing legal, estimating, accounting, or other professional services through this book. If professional services such as these are required, the reader should seek out the services of competent professionals in their area.

Although every effort has been made to have this book be as complete and accurate as possible, there may be mistakes both typographical and in content. This book should be used only to help guide you through the general home building and design process and not as the ultimate reference tool for home building and design information.

The aim of this book is to entertain and educate the reader. Keystone Publishing and the authors of this book shall not have any liability or responsibility to any entity or person with respect to any damage or loss caused, or believed to be caused, indirectly or directly by the advice and information found in this book.

Authors' Story

Building quality custom homes is nothing new for the Cogdill family. For four generations, the Cogdills have applied creative floor plan designs, careful attention to detail, and practical engineering techniques in order to build the best home possible. And today, this family tradition continues with John Cogdill and his son, J.J.'s dedication to designing and building custom homes that have fulfilled American's dreams.

For over thirty-five years, John has been continuously building exclusive custom homes and developing small communities throughout Northeast Florida. After seeing the many mistakes of homebuyers and builders being repeated time and time again, John saw the need for people to become informed on what to expect during the home building process. Unfortunately, when he tried to find a book on the subject, he found that no book existed which could help his clients. Every book on home building or real estate either did not focus on designing and building a custom home or was not written by someone with the practical experience necessary to properly advise his clients. And realizing that misinformation was worse than no information at all, he was forced to inform his clients that there was nothing available to help them.

After several years, John's son, J.J. graduated from the School of Building Construction at the University of Florida and joined the family business. With John's vast construction experience and J.J.'s talent for writing, the father and son duo teamed up to write this much needed book. In it, the writers manage to detail the various steps involved in designing and building a custom home without inundating the reader with technical jargon.

Being one of the only available manuals written by authorities on custom home building, John and J.J. hope that you will read and use the information here to help avoid the many mistakes homebuyers make. With the price of houses climbing every day at an ever-increasing rate, you owe it to yourself and your family to get the accurate information you need to protect your family's biggest investment. This easy to read manual provides you with that information. Now, it is up to you to use it.

The Cogdills' Career Highlights

- Both John and J.J. Hold a Bachelor of Building Construction Degree from the University of Florida

- Certified General Contractor

- Certified Veterans Administration Designer

- Advisory Council, University of Florida School of Building Construction

- Clay County Board of Adjustment and Appeals

- Northeast Florida Builders Association, Past Director

- Clay Builders Council, Past Chairman

- Clay County Builder of the Year in 1990 & 1994

<u>Preface</u>

Over the past forty years, I have seen the home building industry change radically since I first strapped on a tool belt and began to build. Computers have replaced pencil and paper in all facets of construction. Construction managers have had to become more sophisticated to meet the growing demands of our technological times. More information is available for consumption by homebuyers than ever before; yet, confusion and fear about building a home persists. This apprehension by the general public is primarily due to people who write about how to build a home but are not experts in construction themselves. They have never built a home and do not have the professional experience to know whether the information they are publishing is the way the home building process actually works. Therefore, the information they write may or may not be correct, depending on whose opinions they have surveyed.

During the fifteen years I have been teaching here at the University of Florida School of Building Construction, we have been working to stem this tide of misinformation by giving future construction managers the tools they will need to build great futures as well as great structures. To this end we have consistently worked to provide our students and the public with the best information on modern building practices currently available. Therefore, when I received a call from J.J., telling me he and his father had written a book on home building, I was naturally pleased about someone with construction experience and know-how actually writing a book on the subject.

John and J.J. Cogdill are definitely well qualified to write a book on our industry. I had the pleasure of teaching J.J. during his college days at the University of Florida, and found him to be intelligent, a good student, and a man of excellent character. During that time, I also had the opportunity to meet his father John while teaching continuing education courses for professional builders in the Jacksonville area. John and I had many similar home building experiences throughout our years in the home building industry. I identified with many of the stories he describes in this book and even suggested a few of my own. Both John and

J.J. have degrees in Building Construction from the University of Florida, and as you will see from this book, both have used them well.

When I read this book, I was impressed by the fact that although it looks long, it is indeed an easy read. It is a concise, well-written manual that will help both builders and homebuyers alike understand the process of building a home. There are at least two facts on every page that homebuyers and builders should make part of their standard procedures. In addition, the illustrations and highlights at the end of each chapter help to make a subject that can be somewhat confusing, easy to understand and a delight to read.

So if you are going to build a new home, I urge you to read this book. Whether you have built many times before or are about to build your first home, this book will prove to be an invaluable guide for you and your family. And if you are a home builder, I recommend you read this book and adopt the practices John and J.J. have outlined here. Their tried and true methods of home construction will save you a lot of headaches and greatly improve the quality of your homes.

Sincerely,

Dr. Leon Wetherington
Associate Professor
University of Florida, School of Building Construction

Table of Contents

Introduction

For more than thirty years, I have had the greatest job in the world – designing and building custom homes. During that time I have built apartment buildings, offices, warehouses, and done some residential and commercial remodeling work. However, I have always returned home to residential construction.

Building custom homes, I get to work not only as the builder but also as the designer. I meet with the client, find what they need in their new home, help them set a budget, create the structure on paper, and then go into the field and put it together. Just like the builders of old, who spent their lives constructing the great temples and palaces of the ancients, I get to have a voice in the entire design and construction process.

Other types of building, such as commercial and industrial, are designed by a team that handles every part of the design process. This team is usually headed by an architect and his staff along with engineers of various disciplines who handle the details of the overall design. Structural engineers handle the structural system, electrical engineers oversee the electrical systems, and mechanical engineers design the heating and plumbing systems – just to name a few. A commercial builder is limited to only building the structure and usually has little voice in the design of his building. I like being more involved in my work than confining commercial projects allow. That is why I have always found residential construction to be so satisfying.

During the last several years, I have noticed a disturbing trend that has become stronger with each new client. This trend is an increasing anxiety each person brings with them when they ask me to design and build their new home. Instead of being happy and excited about seeing something new created only for them, new homeowners have actually been fearful and apprehensive about their new home. Misconceptions, wrong information, and the "helpful" advice of well-meaning people, who do not understand the designing and building process, have turned buying a new home into something to be feared instead of what it actually should be – something to be enjoyed.

When my son and I decided to write this book, our main goal was to help alleviate some of the anxieties our clients had when they first stepped into our office. We found that the best way to help them get over their fears was to give them the information they needed to make intelligent decisions about designing and building their dream home. After all, taking an idea out of your head and making it into something you can walk through and   actually live in should not be scary. It should be exciting. In fact, planning and building your dream home should be one of the most exciting and rewarding things you ever do. And when you combine the helpful information you find in this book with logic and common sense, it will be.

Now there is one thing I do want to stress before we go any further. It is not our intention in this book to tell you how to build your home. The building process is entirely too complicated to be explained in a book of this size. Our office has literally hundreds of books on construction, and almost every day I still discover new products and techniques they do not contain. Therefore, instead of trying to teach you how to do the job of your builder, we want to give you the information you need to work with your builder and designer to come up with a product that will satisfy your needs. This way, you can be prepared for what you are about to experience and enjoy the process of making your dream home a reality.

Over the past several years I have also noticed that when I try to explain something about the designing/building process to frightened people who do not understand it, they have begun to question whether or not I am telling them the truth. At this point I would like for you to remember that you have not come to me and asked me to build your home. I expect to receive no money when you construct your new home, and I have nothing to gain by covering for another builder. My only goal is to give you the information you and your family need to make informed decisions as you mold your dream into reality.

So I invite you to please, sit back in your recliner, pour a glass of your favorite beverage, and let us discuss the wonderful process of designing and building your new home.

Notes

Chapter 1:
Myths and Monsters

There is probably no activity you will ever be involved in that is as filled with misconceptions and misunderstandings as this business of designing and building a new home. People who do not understand the process pick up bits and pieces, pull them out of context, and create the new home horror stories that you hear passed around water coolers and across back fences throughout the country. These stories cause fear and anxiety and make the process of building your new home much harder than it should ever be.

These tales that strike fear into the hearts of new homeowners are what we call "myths and monsters." They are myths because they are born of half-truths pulled out of context, and they are monsters because they will turn around and bite you. They will not only bite you financially, which alone can spell disaster, but they will also bite you emotionally. These monsters will interfere with your ability to make the logical decisions which ensure that your new home is designed and built the way you want it to be.

I am sure you have heard some of these stories before, and you will undoubtedly recognize several of the situations we are going to talk about in this chapter. Before I can tell you what to expect when you build your new home, we first need to discuss some of these myths and monsters so you will know what not to expect. This way we can start with a clean slate and clear away some of the confusion surrounding the construction of your new home. And, in the process, we will start to replace the **fear** of building your new home with something much more rewarding – the **EXCITEMENT OF BUILDING YOUR NEW HOME.**

MYTH #1:
I have a friend who moved into his new house and the _________ (fill in the blank).

INSERT :   **a) Door fell off the hinges**
                       b) A/C or heat did not work
                       c) Plumbing backed up
                       d) Windows would not open
                       e) Et cetera

Telling horror stories about new homes seems to have replaced baseball as our national pastime. I do not know if these tales are embellished to make them scarier or to make the speaker feel more victimized. But if you take the time to look at them logically, most of them become almost ridiculous.

For example, a fellow builder told me about a customer who repeatedly complained that the cabinet doors in his kitchen kept breaking off their hinges and the drawer guides broke whenever he opened them. The builder told his client that all the drawers and hinges (which had been installed correctly) had been checked on their walk-thru before they closed on the house. This type of cabinet hardware had been used for years without problems in thousands of homes throughout the nation. So there should have been absolutely no reason for the homeowner to have these problems.

During one of his frequent trips to replace the cabinet hardware, the serviceman left one of his tools underneath the counter and returned to the house later that evening to retrieve it. As he walked into the kitchen to find his screwdriver, the serviceman discovered the homeowner's 5-year-old son trying to reach a bag of cookies that his mother had put in one of the top cabinets. Finding that he could not reach them, the child had pulled out the kitchen drawers and made himself a set of "stairs". He had then walked up the stairs, grabbed the bag of cookies, and offered one to the serviceman if he promised "not to tell my mother."

After laughing to himself for a moment, the serviceman showed the owner his son's "stairs" and pointed out that there was no way that the cabinet drawer guides and hinges could withstand the weight of a 5-year-old child. The owner then realized that his cabinet troubles were no more than "child's play" and began to enjoy his new home.

Another common story I hear is about people moving into their house and finding that the A/C or heating does not work. Whenever I hear this my first question is, "Why didn't you check your air-conditioning when you did your walk-thru?" The whole purpose of you and your builder going through your home before the closing is to make sure it is ready to move into and everything is working. Flushing the toilets, opening windows and doors, and starting the heating and A/C unit are all part of making sure your house is ready to live in. If these things are not done, the walk-thru becomes a total waste of time. Small items that could easily have been corrected then become the cataclysmic myths and monsters that are embellished in courtrooms and cocktail parties across the land.

On the other hand, sometimes things do get missed even in the most thorough of walk-thrus. I remember one job we had in a subdivision several years ago where a family lived in their home three or four days and the plumbing backed up. We sent the plumber to repair the problem and he checked the line all the way back to the sewer tap, where the house connected into the sewer system, and found nothing. The plumber then called the utility company, and the service technician discovered that the sewer line going underneath the road had been crushed by a road grader when the road was paved. This crushed pipe was blocking the flow of sewage to the main line and not allowing it to leave the house. This unfortunate incident was not our fault, the plumber's fault, nor the utility company's fault. It was just a freak occurrence that sometimes happens and cannot be helped.

By the way, the above story has only happened to me once in more than 1,000 houses, making it a very infrequent problem, to say the least. In the time it took to find the problem, no damage was done to the house. There was only a minor inconvenience to the owner, service was restored the same day, and everyone lived happily ever after. With the right audience at the right cocktail party, this could become quite the horror story. I just hope if that ever happens, the owner has a lapse in memory and forgets who I am.

Statistically speaking, you are probably as likely to be involved in a car accident on the way to your closing as to have such a freak accident happen to you. A good builder will be conscientious, careful, and build a good product. If you do your homework and find a good builder and drive safely to the closing, everything should work out just fine.

MYTH #2:
The bigger the builder, the better the product.

I cannot tell you how many times I have been to a party and heard someone lament, "I bought from such-and-such builder, and I am having so many problems with my new house. I don't understand it. When my wife and I came into town, we didn't know anybody. So we bought from XYZ because they were the biggest in town. Now I wish I had never heard of them. Every time we call back with a problem, it takes a dozen phone calls to get anyone to come out, three or four trips before anything is done, and it has just been a real nightmare to get the least little problem taken care of."

Situations like this happen because a lot of people make the mistake of thinking bigger equals better. They believe this not because of any psychological theories by Sigmund Freud, but because they think that builders who sell the most homes per year must be better than their smaller volume counterparts or they would not sell that many houses. Unfortunately, the total number of homes a builder sells per year has **<u>ABSOLUTELY NOTHING</u>** to do with the quality of his work.

Some of the finest homes in our area are built by builders who actually construct **<u>less than 25</u>** homes per year. And on the other hand, some larger volume builders also do an outstanding job of producing quality homes at a good price. Quality is not a question of volume. Quality is a question of training, experience, and the dedication of your builder to making sure that his work is of the highest quality possible.

Professional, competent supervision during the building of your new home is the only effective means of making sure that your home is built the way it should be. The ability of the people supervising the construction work is directly proportional to the quality of the work done on your new home. When you have highly trained, dedicated people with years of construction experience supervising the construction process, your new home will be a quality product and serve you well in the years to come. On the other hand, when you have poorly trained, uncaring people with little experience overseeing the building process, your new home will be a poor product and give you nothing but headaches and real horror stories to tell your friends.

Unfortunately, many building companies both large and small have a tendency to concentrate on volume rather than quality. These companies work with their sales and marketing teams more closely than they work with their design and quality control personnel. And although they may produce a large number of homes, they may also produce some of the poorer quality homes that you will occasionally find.

An additional reason why volume is always a secondary consideration when choosing your builder is the harsh reality that the average home in the United States will far outlast the company that built it. Because the life span of a construction company is that short, many people make the mistake of thinking that a small volume builder will not survive in the marketplace.

In reality, many of the builders you find that have been active in the local market for the longest period are smaller volume builders. They use a "hands on" management style, allowing them to control costs and run a more efficient operation by having less layers of management than larger-volume competitors. Small builders can also control their volume in such a way that they do not get caught with a large number of unsold homes during periods of economic decline. This small inventory of unsold homes allows smaller builders to weather the ups and downs of the economy, and sometimes makes them more stable than larger volume competitors.

Most large-volume home building companies are run like other large corporations in other industries. They are headed by a Chief Executive Officer or CEO who answers to the stockholders of the company. These stockholders are people who have invested their money in the company and want to see a profitable return on that investment. In order to maximize corporate profits, these large volume builders often increase their volume during an economic boom time to the point that when the market slows down or goes bust, they cannot slow down or stop building fast enough. As a result, the large builders wind up with a large inventory of homes that cannot be sold. And if this standing inventory of homes is too large, it can, and often does, force a company into bankruptcy.

Some people fall under the delusion that many of these large volume builders are so big that they cannot go under. You have only to investigate other similarly sized companies listed on the New York Stock Exchange that have been in financial trouble over the last several years, particularly during the recession of 1990, to see the true story. One look will show you that there is no such thing as a company that is too big to get into financial trouble.

So once again, the number of homes a company builds per year has nothing to do with the quality of your home or whether your builder will still be in business several years down the road. It is, and will always be, the training, experience, and dedication of the supervisory personnel who actually build your new home that will determine the quality of your new home.

MYTH #3:
"I Don't Believe the Mortgage Company/Banker/Developer Would Allow an Incompetent Builder to Build in This Community"

For years I thought that a mortgage company would want to do business only with builders who never went broke. I believed this because if I were running a mortgage company, the last thing I would want is to get stuck finishing the products of a bankrupt builder and have to unload them to recoup my investment. I would put my company's money in builders whom I knew would not leave me holding the bag and whom my customers could count on if any problems came up. I thought this way for a number of years before I realized that economics makes some mortgage companies think exactly the opposite way.

The key thing a mortgage company looks for in a builder is not necessarily the quality of his products, but the number of homes he builds. Volume is the key, not quality. You see, a mortgage loan officer gets paid by the number of loans she makes – the greater the number of loans, the more she gets paid. If she gets 100 houses per year from a builder, and the builder goes broke at the end of five years, the mortgage company may wind up having to complete ten or twenty houses. But during those five years the builder was in business, the mortgage company handled over 500 mortgage loans. During those same five years a small volume builder, who never goes broke and builds twenty houses per year, may leave the mortgage company with no houses to finish, but he only gives them 100 loans to process. And since the mortgage company would rather be paid for 500 loans than 100 loans, they will often prefer to do business with the larger volume builder.

The same kind of high volume thinking also occurs from a banker's point of view. Like mortgage loan officers, bankers also get paid by the volume of construction loans. The more loans they close, the more fees they collect, and the more money they make. Therefore, bankers may be more concerned with volume than the quality of their builders' homes.

Developers also want in on the high volume action because the developer gets paid by the number of lots he sells in a given area. The more lots he sells, the more money he makes. And the faster he can sell his lots, the faster he can repay his loan from the bank for developing the land, and the less interest on the loan he has to pay. Since the developer gets paid regardless of the quality of the homes built in his subdivision, he is eager to do business with big volume builders, no matter what kind of buildings they throw together.

A perfect example of this happened about a year ago when I called a local land developer and asked about his new subdivision.

The developer said he had sold all his lots to a local builder who had a reputation for building small, low income, low quality housing. When I asked the developer why, he simply replied, "Because I know he can "take down" (buy) the lots fast." And who could argue with logic like that? With an acquisition and development loan (A&D Loan)[i] of over $1,000,000 and a high interest rate to pay, fast sales for this developer were not only desirable, they were essential.

MYTH #4:
"They Don't Build'em Like They Used To"

From the time I was a little kid, all I ever heard people say when something new came along was, "They don't build'em like they used to." For some reason, most people believe that if a building has been around a long time, it is better than what is built today. For those people I have a simple response; "Take a closer look."

If you believe older is always better, then think about the gigantic stone castles of Europe that were built during the Middle Ages. These massive structures have been standing for well over 500 years, so they must be better than what is being built today, right? Well, consider that when they were built there were no modern building codes to govern their design. Without these codes to guide them, ancient builders designed in such mistakes as stairs with no railings and step risers and treads of such length that visitors tripped and fell when they used them. And when you take into account the lack of heat, indoor plumbing, ventilation, and insulation, you can easily see that living in these monstrosities would be intolerable. Cold in winter and hot in summer, these dark, musty, dreary places allowed residents to take up new wintertime activities such as huddling around the fire to keep from freezing to death and watching water seep through their six-foot thick stone masonry walls.

Is this a better way to live than in a modern, climate controlled home? I personally do not think so. But if you are one of those who thinks older buildings are always better, please feel free to move right into one. But do not expect many of your friends or family to stay long. (Although with some family members, this <u>may</u> be something you want to look into.)

MYTH #5:
"How Much Do You Charge Per Square Foot To Build A House?"

This question has caused more home buyers to part with their hard-earned savings than any other question I have ever heard. And the reason this question is so dangerous to a new homebuyer is because most people:

a) Do not understand how to compare the cost of new homes, or

b) Do not understand how builders determine the price of a new home

For some reason, many people have the preconceived notion that somewhere in the universe there is a mystical price per square foot that all builders in the world know and use to price their model homes. Therefore, if you can take the sales price of a builder's model home and divide it by its respective square footage to come up with a price per square foot, you can use this price to drive down the price of another builder's model across town. After using this procedure with several builders, you will eventually bring the price down to where you get the lowest price per square foot and the best value for your money.

This sounds great in theory. Unfortunately, when you try to negotiate a bargain using this price per square foot method out in the real world, chances are the only thing you will wind up negotiating is a complete mess.

Comparing two houses by their cost per square foot is like comparing two cars by their cost per tire. If you have a small economy car and compare it with a large luxury model, both of them have four tires. But when you divide their sale prices by the number of tires (4 for each car), the price per tire varies significantly.

The same holds true for a house. If you have a three-story Tudor mansion with marble floors and mahogany paneled walls, it will cost more when you divide the sales price by its square footage than a less imposing starter home. When you consider the workmanship and the features that come with each house, the three-story mansion is obviously the place you would rather live. But when you compare to two houses only by their price per square foot, it seems like the simpler, less imposing structure is the one you would rather call home.

I know some of you are thinking, "But these are such extremely different houses; of course it doesn't work. It only works when the houses are similar to each other." Well, the truth is that the cost per square foot method fails here too for precisely the same reason. You see, I can take the exact same house with the exact same floorplan, the exact same lot, and the exact same square footage and keep adding options to one house until I have **doubled** the price of the home **without changing the square footage at all.**

The list of items that can change your price per square foot without changing the square footage could fill a book in itself. Two of the more obvious of these changes is whether or not the sales price includes the price of the homesite and the closing costs. Also, the allowances[ii] which the builder has included can be drastically different, making the sales price of one home much less than the other. For example, in our standard homes up to 2000 square feet (SF), the owner is usually given a lighting fixture allowance of $800. This means that the owner can spend up to $800 on lighting fixtures for his new home. If he spends less than the $800, he is refunded the difference; if he spends more than $800, he pays the difference.

One allowance which can make a monumental difference in the sale price of your home is landscaping. The landscaping allowance for a given home can vary anywhere from 300-400%, depending on how much landscaping you include with your sales price. Do you sod the entire front and back yard, or do you sod the front yard and sprig the back? How much shrubbery do you include? On a typical starter home, you can get by for as little as $100, while on a more expensive home, the shrubbery can easily cost anywhere from $5,000-10,000. Sprinkler systems, rock gardens, elaborate fishing ponds, and even a swimming pool can be included in a landscaping allowance. These extra goodies can make the sales price of your home shoot up anywhere from $1,000 to $40,000 or $50,000 or more without changing the square footage of your home one bit.

When you start looking at the interior of your home, the finish flooring and appliance allowances can change your total sales price by as much as **ten percent by themselves.** Ceramic tile flooring will usually run about four times more than carpet or vinyl flooring for the same area. Vinyl flooring can vary in grade or quality from inexpensive to expensive grade that costs four to five times as much as the vinyl which would be included in a starter home.

Cabinets are another allowance that can make your sale price shoot through the roof. Kitchen cabinets can vary as much as 300-400% in price, depending on the quality and the number of cabinets you include. For example, you may include an island or a counter bar – items which may add $1,000-1,500 or more to the price of your home without affecting your square footage. And when you throw in such bath fixtures as extra sinks and vanities, whirlpool tubs, and maybe even an extra bath or ½ bath, you drastically change the sales price without augmenting the square footage.

And while we are on the subject of the home interior, you must include the electrical system and the countless options that can be included there. Does your system have a higher rated electrical service[iii] so that it can handle more current being drawn through your home? If it does, then you can start adding electrical features such as: dedicated circuits[iv] for computers and microwaves, extra lighting, extra electrical outlets, extra freezers and refrigerators, and countless other costly electrical conveniences that make your life more comfortable.

As you can see from this short list of items, it is very easy to either double the price per square foot or cut it in half from one identical home to another. When you look only at the price per square foot without comparing all of the items included in a home, you have much less information than you need to make an informed decision. Therefore, to effectively compare the price of two similar models built by two different builders, you must look at all of the features and allowances each builder has included in their sales price and base your comparison on the total package.

At this point, some of you are probably asking yourselves, "If the price per square foot method is so ineffective, then why is it such a popularly-held belief?" The method of pricing houses per square foot is a case in point where people who do not understand a process pull something out of context and distort its meaning. The square foot pricing method is often used by statisticians to compare the prices on homes in different regions of a certain state or across the country. For example, a statistician might say that homes in Santa Barbara, California, are selling for an average of $195 per square foot while homes in Jacksonville, Florida, are selling for an average of $55 per square foot. Therefore, you can expect a typical home in Jacksonville to cost one-third to one-quarter the price of a similar home in Santa Barbara.

In addition to statisticians, appraisers also use the price per square foot method to establish the most reasonable market value for a proposed new home. However, they do not just arbitrarily compare the price per square foot on homes that have recently sold. An appraiser has a tremendous amount of training in how to compare a home with others that are similar but not identical. He has a set procedure for comparing the differences in these homes and establishing values for them so that a reasonable market value can be obtained. First, the appraiser acquires a set of plans for the home he is appraising (the subject property), determines the square footage, how the home is constructed (brick, concrete block, wood frame, etc), and finds any other features the home contains which will add to its value. He will then find comparable homes in the area (comps) which have sold recently, and will research those homes thoroughly. The appraiser will start his research in the public records and find the sales price of his comp homes. Then he will call the sellers and/or the builder and ask them a number of questions about these comp homes. Often the appraiser will even get a copy of the comparable homes' plan and determine some of the features the comp has that add to its value. Once the appraiser has a list of these items, he will go through a long, elaborate process of assigning values to each feature and adding or subtracting these values to and from the comp home to make it identical to the subject property. The appraiser then takes this adjusted value and calculates a price per square foot to use in his evaluation of the subject property.

In most cases, a person who is not trained to do appraisal work simply does not have the information necessary to adjust these values and come up with an accurate cost per square foot. Therefore, this is not an accurate method for the lay person who is trying to figure out what they can afford because they simply do not have the training, experience, or data necessary to use the price per square foot process correctly.

Instead of taking the sales price of new homes and dividing by their respective square footages, it is far more effective and much easier to simply take homes which are within a hundred or so square feet of each other in comparable areas of town and compare them based on their sales price and included features. Take a notepad with you to each home and make a list of the items included and the sales price. After reviewing your list, you may be surprised to find that a 1,900 square foot house on a better lot with a better floorplan and nicer amenities, such as upgraded appliances and higher ceilings, is a much better buy than a 2,000 square foot house with fewer of these amenities.

So when you compare similar homes, if you worry less about square footage and concentrate more on the floor plan, room sizes, and what is included with the sales price, you will have a much greater chance of finding the best home for you and your family.

MYTH #6:
"I'll check the job and make sure it goes right"

A couple of months ago I was having lunch with my banker, and she was laughing about a local builder who had to resubmit some paperwork to her mortgage department. It seemed that when the builder poured the concrete slab on a custom home, he was off with the slab measurements by more than two feet (1/2" is the maximum tolerance allowed on most jobs, including ours). When the error was discovered a week later, the builder corrected his mistake by hacking off part of the slab and called in a structural engineer to design extra footers[v] so the slab would still support the house. When the owner drove up and saw the jagged edge of the slab where finished concrete had been the day before, he forced the builder to start his house over again on a neighboring lot and resubmit the mortgage paperwork to the bank. The builder then finished the house and resold it to another family who had not seen the broken slab supporting their new home.

When I heard this, my first reaction was, "If my builder had screwed up that much on the foundation, what else would he screw up that I didn't know about?" As a potential homeowner, you have the right to assume that someone presenting themselves as a builder knows what they are doing and will not allow potentially large mistakes like this to happen. Although such mistakes are rare, you as a homeowner cannot and should not be expected to catch your builder's foul-ups. You are not the job site superintendent, and chances are you will not be out there everyday checking the job. And even when you do make it out to see your house, how will you know what to look for? Do you know how to tell a big construction problem from something that is not important? Very few homeowners can. Most people do not have the training and experience to know what to look for any more than I could do brain surgery. Your builder is the construction expert; therefore, you should not have to be an expert in construction yourself. If you try to supervise your home without having adequate construction training and experience, you will only create problems, not alleviate them.

Every time I hear someone say they can supervise the job themselves, I think about a home I built for a young couple many years ago. The husband was a Navy officer and the wife was an operating room nurse – very intelligent people, who had carefully compared prices and chose us to build their home. Things went well during construction, and when it came time for the final walk-thru the husband was sent on deployment with the Navy. Therefore, I did the walk-thru with only the wife. After I gave her my standard speech of how we were going to go through the home room by room and check everything, she told me, "This is our first home, and I have never done a walk-thru before. So I'll have to rely on you to explain what we need to do."

Sensing her apprehension, I tried to make her laugh by saying, "You know, a long time ago I realized something about the medical profession."

"What's that?" she asked.

"This germ theory business you guys have been promoting all these years is not real at all. It is just something you invented to keep the lay man out of the operating room."

She laughed and quickly retorted, "You know, you just may be right. But it's no different than builders who have their superintendents chew tobacco so their buyers will be repulsed and not ask too many questions."

"Yeah, but your method actually works," I countered.

Unfortunately, tobacco-chewing superintendents are not nearly as effective as the germ theory in keeping the layman out of our construction operating room. Therefore, you, as our client, simply must understand that the reason you are hiring a builder is because you do not have the knowledge to build your home yourself and do not have time to learn. If you could do it yourself, you would not need a builder. You could instead pocket the money you are giving him to oversee your new home's construction and spend it on something else. If you could not think of anything, I am sure you could find a number of friends to offer a helping hand. After all, what are friends for?

Instead of supervising the job and making sure the work goes according to schedule, your primary concern should be picking a builder who knows what he is doing and will put supervisory personnel on the job who will make sure your home is built properly. Now this is not to say that if you have a question you should not ask it. Our most informed buyers have many questions during the course of construction. Therefore, if something comes up that you need to ask your builder, pick up the phone and ask him about whatever you saw that may be a source of concern. And remember, the main thing you should do when you ask your builder a question is **listen** to the answer he gives you. Too many times people have the tendency to listen solely to friends and relatives who have no training in construction. These people are well-meaning, but tend to fill buyers' heads with those famous builder horror stories we mentioned earlier. Our clients then come to us scared to death and filled with preconceived ideas on what the answers to their questions should be. And when we do not give them the answers they expect to hear, their fear is magnified and reinforced. This fear does nothing but cause needless friction between you and your builder and hamper your builder's ability to make sure that your home is built the way it should be.

So when you have a question, ask your builder. And if you have a follow-up question, ask him that question. But make sure that you <u>listen</u> to the answer he gives you, and if it is not the answer you were expecting, ask him to explain his answer in greater detail. Most builders, including myself, would rather take an extra ten or twenty minutes out of their busy schedules to make sure you understand what is happening than let a misconception fester to the point where it becomes a problem. And if you have done a proper job selecting your builder, the answers you get from him will be truthful and correct and leave you with no lingering doubts about the quality of your new home.

MYTH #7:
"I'll serve as my own builder and subcontract the work out and save at least a third"

I wish that I had made as much profit over the years as people think I did, because then I could have stopped working twenty years ago. Most people seem to think that a builder will make at least a third on each house he builds. If that were true, then on a $200,000 house the builder would earn a little over $65,000. Unfortunately, builders usually make less than that – much less. On average a residential builder usually makes about twelve to eighteen percent gross profit on each home he builds. Then once company overhead is subtracted for insurance, employee salaries, equipment, taxes, and other costs of doing business, the builder might end up keeping about five to seven percent[vi] of that profit if he has a good year. And if he has a bad year, the builder gets to keep nothing but the hope of better years to come. These low profit margins are the reason why a builder has to be a good manager if he expects to keep food on the table.

One of the main ingredients in being a good construction manager is knowing which subcontractors to hire and what the price of their specialty work should be. When choosing a subcontractor, a low price is not necessarily the most important factor to consider. A good subcontractor must have the ability and manpower to do his job, and he must be financially stable so he will still be in business if you have a problem. There is nothing worse for the homeowner and the builder than to have something like an air-conditioning unit installed in a home and find a month later that the system was installed improperly by a subcontractor who has since left town. Because when the subcontractor is not around to take care of a problem like this, the owner and the builder are left holding what can sometimes be a very heavy bag.

Our plumber came in the other day and told us about a guy he hired to start working for him and fired three days later. He canned the guy because he was so efficient – he did more damage in one day than his other men could fix in three. When the worker defended himself by saying, "I can get this right if I have a little more time," the plumber replied, "Son, I'll run out of money before you run out of time. And I can't afford it."

Just like our plumber, builders run into "efficient" subcontractors that waste more time and money than we ever thought possible. This is what we repeatedly try to relate to people who have no construction experience but are convinced they can build their own house. All you have to do is pick **one** wrong subcontractor and he can wipe out **all** the money you save in a house by doing the work yourself.

If you are able to build a perfect home yourself, you can reasonably expect to save about $1,000 on each $10,000 it costs to build your house. This translates into about $20,000 in savings on a $200,000 house. When you consider that there are at least twenty or more subcontractors involved in the building of your new house, you can see that there is a great potential for one or more of them to make a costly mistake. And when you consider the fact that in over thirty years and 1,000 houses I have **never** seen a **perfect** home, the idea of building your own should make even the biggest Scrooge run for the nearest qualified contractor.

MYTH #8:

"If I buy my own lot and have the builder build on it, I will save money."

Over the years, we have constructed many homes both in our own subdivisions and off-site on clients' lots. In almost every case, it cost the homeowner more to build his home on land he already owned. This was not because I felt like charging him more money. It was because there are extra expenses that go along with building on your own lot.

These extra expenses can include well and septic tank installation instead of water and sewer hook-up, extra landscaping and fill dirt, and extra engineering costs that can come from the makeup of the subsoil or the topography of the land. Also, the builder is usually set up to construct multiple houses in a subdivision instead of doing only one on your lot. This means that labor and material charges are less for him to build in a subdivision, and he will charge you less if you build there. This is why in most cases you can actually save money if you build in a subdivision instead of on your own individual lot.

MYTH #9:

THE ___________ WILL PROTECT ME.

INSERT:
      a) Building inspector
      b) Veterans Administration (VA)
      c) Mortgage Company
      d) Realtor®
      e) Lawyer
      f) Deed Restrictions

Several years ago I did a routine check on a concrete slab that was being prepared for one of our homes. When I saw how deep the bearing footings[vii] had been dug, I found that instead of being twelve inches deep like they were supposed to be, the footings had been dug maybe one or two inches. When I asked one of the men preparing the slab why they were so shallow he replied, "The inspector said it was OK." Since my company name was hanging on the sign in front of the job, I gave him a choice: he could either dig the footings as deep as I wanted them, or he could have the inspector sign his paycheck. The footings were dug to the proper depth.

In most jurisdictions, the building inspection department is composed of a chief building official who has field inspectors under him to make sure that new buildings meet all the building codes. Most of these inspectors do a good job and should be commended for the work they do. However, given the thousands of pieces and parts of a house and the encyclopedic volumes of codes that apply to new construction, it is impossible for an inspector to spend the extremely short amount of time he spends on the job and find every problem that might exist. Inevitably, things such as the footings mentioned above may slip through the cracks without adequate builder supervision.

Any professional building inspector will tell you that no amount of inspection can replace a qualified superintendent working with qualified tradesmen on the job. A building inspector serves as a backup to double check a job, not as the primary construction supervisor. He simply does not have the time to watch every nail driven and every brick laid. Therefore, the building inspector helps the supervisor all he can by trying to catch what the supervisor may have overlooked and make the job run as smoothly as possible. He and the superintendent work as a team, because no building inspector can be expected to do this tough job alone. Only by working together can a building inspector and a qualified superintendent make sure your home will stand the test of time.

The same thing holds true for a Realtor® or site agent. These people also do not have time to supervise every construction activity that happens with your home. Not only are they on the job site less than a building inspector, but they usually do not have the same in-depth construction background as an inspector. Realtors® and site agents are trained to locate property, keep up with marketing trends, know the market value of property, know which areas are growing and declining, and how to obtain financing for home mortgages. These people know their jobs and do them well; however, their job description does not include supervising the construction of your home. Therefore, you still must have a qualified superintendent overseeing the work.

Some veterans believe that the Veterans Administration (VA) from which they can obtain mortgages will protect and ensure the quality of their new home. While it is true that VA sets certain standards of compliance for new homes, most VA inspectors today no longer inspect your home while it is under construction to make sure these codes are being met. Instead, they rely on local building departments and only do a final inspection after the house is complete and ready for move-in. During this final inspection, the structure is covered inside and out with finishing material such as stucco, gypsum wall board, or brick. Therefore, the inspector cannot see the structural system. And if he cannot see the structure, how can he possibly make sure that your home meets <u>all</u> the code requirements?

I know that by this point some of you are asking, "What about the pages and pages of deed restrictions that govern my new home community?" Well, I am glad you brought that up because these can be misleading. The truth about deed restrictions is that they are sometimes enforced by - **NOBODY**. In theory, they are supposed to be enforced by the developer while the subdivision is being built. Then when all the homes in the subdivision are completed, this chore is handed over to the local Homeowners' Association.

The Homeowners' Association is made up of all the homeowners in a community who theoretically make sure the deed restrictions are met by all the residents of the area. In reality, these organizations often end up with little power and only enforce the rules sporadically. Usually the offender has to be taken to court and the association has to pay expensive legal fees. Since many associations simply do not have this kind of money, the rules are often difficult, if not impossible to enforce.

During the recession of 1990, the developer of a local high-end subdivision was facing bankruptcy. Ordinarily this developer was very strict on enforcing deed restrictions and maintaining the integrity of his finished product. But as you can imagine, impending bankruptcy is one of the greatest behavior modifiers known to man. It can make someone do things they would not normally do. And in this case, it made the developer sell a group of lots at a greatly reduced price to a builder who specialized in lower-cost, tract-type housing.

As you can probably guess, this caused quite an uproar in the area. The homeowners felt that when this new builder came into their community, the quality of new homes being built would decline and many of the restrictions governing them would simply be forgotten. Well, the truth is that when the builder moved into the area, that is exactly what happened. And unfortunately, there was very little anyone could do about it.

For a while, the cost of new homes being built in the area were substantially less than what had been built at the start of the development. The look of the area did suffer some decline, and those who decided to sell their homes suffered a tremendous loss of equity in their investment. However, those that stayed and decided to ride out the storm were rewarded when the market began to recover and the subdivision reverted to higher-end housing. Their property values went up, but they never quite recovered from the weight of the lower-end houses that drag them down to this day.

As a last resort, many people figure that their lawyer will take care of them through anything – including the construction of their new home. Unfortunately, this is another of those misconceptions that cause people to have a false sense of security when their home is being constructed. A lawyer is trained to be familiar with the laws that govern the people in the state where she practices. Nowhere in her training does a lawyer learn how to construct a building or what to look for when she steps onto a construction site. Therefore, your lawyer will not be able to ensure that your builder is doing his job correctly when he begins building your new home.

The truth is that a lawyer does perform many extremely important functions during the construction of a new home. These functions include the following:

- ➤ checking the title
- ➤ writing and reviewing real estate contracts
- ➤ closing the construction loan
- ➤ checking liens against the property
- ➤ recording the deed
- ➤ reviewing and recording the mortgage
- ➤ generally insuring that their clients' legal rights are protected

Since the laws vary from state to state, the services of a lawyer are very worthwhile, even if you are familiar with real estate contracts.

Some people feel reluctant to hire a lawyer because they want to save money. What they do not realize is many thousands of dollars would be saved if they allowed a lawyer to look over their contracts before entering into them. A lawyer is available to solve problems – not create problems. Many times over the years we have provided blank copies of our contracts to buyers so that their lawyers could look them over before they were signed. In addition, I have also met with several clients and their attorneys to go over portions of our contract and work out any provisions which may have needed to be altered for that particular client. No legitimate builder would ever balk at having an attorney go over his contract before it is signed. Instead, a legitimate builder would prefer to have a lawyer ask him questions so that any potential misconceptions can be cleared up from the outset.

Protecting your rights under the law is what you are paying your lawyer to do and what you should expect from her. However, a lawyer is not trained in construction and therefore she will not supervise the construction of your new home. That is the job of your builder. So make sure your builder is doing his job by having a qualified superintendent overseeing your home's construction. Because if your builder is doing his job properly, you will not have to hope that someone is ensuring the quality of your new home. You will know that quality is included with your sales price.

MYTH #10:

"I'll sue! I'll sue!" (But will you collect?)

Over the years, I have had a few people who thought the quality of a building was directly proportional to the builders' fear of lawsuits. The idea that someone actually thinks that they can frighten a person into doing their best work seems ludicrous to me. I have always found that such a climate actually fosters the exact opposite – a desire to do the least amount of work possible to get through with the job and get away quickly. Besides, the truth is that if the construction of a structure is bad enough to sue over, and you go through the long, windy process of a lawsuit and wind up with a judgement against the builder, the chances are that by then the builder is broke, and any assets he may have held have long since disappeared. And by the time you get a judgement against him, several others have too, and the builder and any money he may have had have vanished into thin air.

I knew one material supplier in town who had so many liens against local builders for failure to pay their bills that he wallpapered the back of his office with them. After trying to collect for months and not getting anything but a big legal fee for his efforts, he decided to stick them up on his wall as a conversation piece. That is until his wife decided they did not match the rest of the office décor and forced him to take them down.

It should never come to this. If you get to the point that you have to threaten your builder with a lawsuit, your relationship has probably deteriorated so much that neither of you will ever be satisfied with the finished product.

Having served as an expert witness in several lawsuits where owners were suing their builders, I can tell you firsthand that very little good ever comes from these types of proceedings. Coming between a builder and an owner engaged in a lawsuit is like trying to separate two Dobermans in a dogfight – one of them is going to bite you when you are not looking. Usually when the building process has degraded to the point of a lawsuit, the two parties are looking more for blood than justice.

What has always fascinated me about these lawsuits is the fact that all of my credentials and all my years of experience matter most to someone who wants me to testify in their behalf against another builder. When people are trying to find a builder to construct their home, my credentials and experience are usually given no more than a casual glance. But when it comes to suing the builder they picked to build their home, all of a sudden my qualifications are very important to them. If only they had thought to do some research and find a qualified, experienced builder when they were building their home, they would never have had to convince me to testify. The job would have been done correctly in the first place. And isn't it easier to make sure your home is done right from the start? The court system in our country is already overcrowded with frivolous lawsuits. Why go through the process when you can so easily avoid it with a little advanced preparation?

So, if you find yourself in a lawsuit with your builder, then you have made a mistake somewhere and did not follow all of the advice in this book. So please, read carefully because there is a lot more to learn before you start designing and building your dream home.

Highlights of Chapter 1

- Most new home horror stories you hear are almost ridiculous when looked at logically.

- The total number of homes your builder sells per year has **ABSOLUTELY NOTHING** to do with the quality of his work.

- Professional, trained, competent, and dedicated supervisory personnel directly overseeing your home's construction is the **only** way to guarantee your home will be well-built.

- Mortgage companies, bankers, and real estate developers are more concerned with the number of homes a builder constructs than the quality of his work. Because the more homes a builder builds, the more money they make.

- Modern, climate-controlled homes are better than older homes because they have modern building codes to govern their design and construction.

- Comparison shopping for homes should **NEVER** be done by comparing their price per square foot. Instead, take similar homes within 100 square feet of each other in comparable areas of town and compare them by their sales prices and included features.

- Instead of supervising the job, your primary concern as a homebuyer is picking a builder who knows what he is doing and will place competent supervisory personnel in charge of your home's construction.

- If something comes up that you need to ask your builder, pick up the phone and ask him. **Listen** to the answer he gives you and ask him to elaborate if his answer is not what you expected.

- Building your home yourself, you will be able to save money if everything is built _perfectly_. However, keep in mind that in over 30 years and 1,000 homes I have _never_ seen a perfect home (and never heard of one either).

- Building on your own lot will cost more, not less, than building in a subdivision.

- Your local building inspectors, Veterans Administration, mortgage company, Realtor®, lawyer, and deed restrictions cannot ensure the quality of your home because they are not overseeing your home's construction everyday.

- Research local builders and their **qualifications** carefully before choosing one to construct your new home. If you do not and end up suing your builder, you will probably not receive any money since others will most likely have received judgements against him before you.

[i]   A&D Loan – the money a developer borrows from the bank to develop land. He later repays this loan with interest from the sale of lots in his development.

[ii]   Allowances- a specific amount of money set aside in a construction contract for certain categories of construction items (i.e. lighting fixtures, appliances, landscaping, etc)

[iii]   electrical service - the amount of amps which can be drawn through your home's circuit breaker panel without causing an electrical short. The higher your home's panel amp rating, the more electrical items you can run off that panel without causing an electrical problem.

[iv]   dedicated circuit – electrical outlet which is the sole outlet on an electrical breaker in your electrical panel box. Allows you to plug in an appliance which takes up a lot of power without fear of tripping your electrical breaker.

[v]   footers – deepened sections of concrete with reinforcing steel in a concrete slab. Used to support heavy loads and structural walls to allow the building to support itself.

[vi]   <u>Builder</u> magazine, May 1998, "Dream Market Has Builders Wondering: How Much Longer?" pages 88,128.

[vii]   Bearing footings – see concrete footers.

Notes

Notes

Chapter 2:
The Bid Process – the Biggest Myth Of All

"I was low bidder on every contract."

Comparing costs on similar goods so that you pay the lowest possible price is one of the first things you learn to help make your money go farther. When you go to the grocery store, shop for a car, or even look for that special doll your daughter wants for Christmas, you always shop for the lowest possible price. So naturally when you start looking for a home, you try to use the same cost comparing techniques to get the most house for your money. This shopping method is known as the bid system.

Getting bids is one of the oldest ways of comparing prices on construction work known to man. Used as early as the time of the ancient ruler Hammurabi in the eighteenth century BC, the theory states that when you give a number of qualified builders the exact same plans and specifications, they will give you a price based on identical work. The qualified builder that has the lowest price is then hired to build the building.

That is how it is supposed to work, in theory. Unfortunately, the bid process almost never works correctly because most people do not realize how much detail a bid must contain and cannot properly evaluate bids. The perfect example of this happened in 1990 when a man called me who had drawn his own house plan and was getting bids from several builders. Now normally when I get a phone call like this I avoid meeting with them because the plans they have drawn themselves are never specific enough to come up with an accurate bid (you will see why I say this later). But he happened to call me in 1990, when the local housing market was in a serious decline. Neither I nor any other builder in town at the time could sell anything to anybody, anywhere, under any circumstances. These bad periods strike builders many times over our careers, and when they do come, the only thing we can do is ride them out and get work wherever we can.

So when I received his call, we made an appointment, and he came by with a "plan" similar to the one you see on the next page. When I asked him where the rest of his plan was, he responded, "Well, your competitor down the street gave me a price on this. Are you telling me that you're not as good as he is?"

I knew right then I would not get this job because I have seen this scenario play itself out so many times before. The plan you see here is so vague that it could cost anywhere from $200,000 to $2 million to build. The builders who give him a price will adjust the room sizes to whatever looks reasonable, put in the finish flooring they think looks best, make the outside of the house look like whatever they prefer to build, and give the buyer a price based on their specifications. And since no two builders will have the exact same specifications, each price will be based on a different house – a fact the buyer will not realize when he chooses his builder.

The buyer will then pick the builder with the lowest price and think he got an incredible bargain. After congratulating himself for being such an astute businessman, he will do one of two things:

1) wander away for several months and come back to find a house completely different from the one he wanted or,

2) stay on the job and fight the builder each and every day.

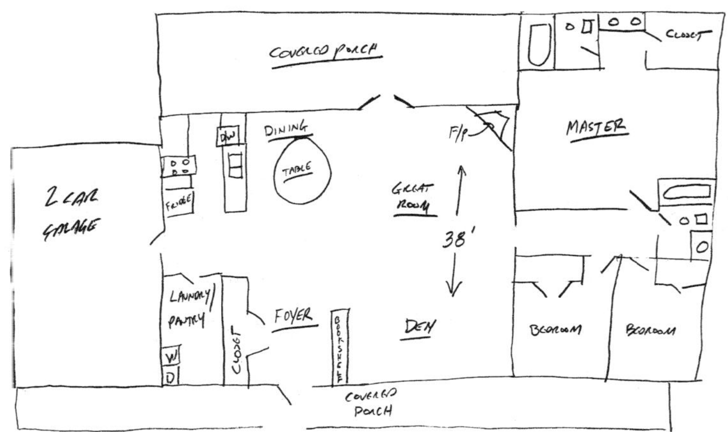

<u>An incomplete bid package</u>

Either scenario is a nightmare for everyone concerned because the result is always messy.

The buyer will rant and rave about all that is wrong with his house, while the builder will tell him repeatedly that the house plans were approved by the buyer, and he is sticking to the approved plans. The buyer will then complain to every governmental agency that will listen and have the local news media film his house and all that is "wrong" with it. And when all is said and done, the buyer will never be happy with his new home, no matter what the builder does to appease him.

"How can I avoid having this happen to me?" you ask. Well, the best way to avoid a similar mishap is to understand the bid process. First, a complete bid package will have a full set of plans and specifications that cover all materials and methods that will be used to construct your home. Second, the allowances the builder gives for appliances, finish flooring, and so forth will be stated and the same amount included in each bid. And finally and most importantly, there must be a designer/architect who designed the new home available to answer any questions the bidding builders may have while they are estimating and bidding the job. The fee charged by this professional will probably be very high, but the monetary and emotional damage he will prevent shall prove invaluable. To fully understand just how much detail is involved in an effective bid process, we have included a bid package for one of our model homes which contains the bare minimum an architect or design professional needs to solicit bids for your new home.

First, the bid package must contain a complete set of plans and specifications like you will find at the end of this chapter. The plans tell you what your new home will look like and include:

Plot Plan	shows the placement of the house on the lot; design driveway, walkways, and patios; dimensions of home; distances/setbacks from property lines; length of property lines; bearing of property lines; elevations of slab & property corners; drainage type with arrows showing water runoff direction; legal description of the property; and the North arrow
Foundation Plan	tells what kind of foundation your home will have; placement of reinforcing steel; size of bearing footings; wind anchorage; etc.
Floor Plan	shows placement of walls; wall sizes; ceiling heights; cabinet placements and layouts; finish floor (carpet, tile, vinyl, etc) placement;

interior details; and includes notes from the specifications on finishes, plumbing, windows, trim finishes, electric, and heating and air-conditioning; structural details – beam and column locations, sizes; special notes

Elevations

shows how home will appear from exterior on: left, right, front, and rear; exterior finish material (stucco, brick veneer, or wood siding); roof slopes; shingle type; and any exterior architectural details (stucco bands, brick quoins, etc.)

Electrical Plan

displays outlet placement; light fixture placement; ceiling fans; doorbell chimes; A/C compressor size; phone jacks; cable TV jacks; special light types and placements (recessed, bubble, etc); cover plate color; amperage of the service box supplying the home

Roof Framing Plan

shows placement of roof trusses and outline of hips and valleys of the roof from a view above the home; includes a guide for framers in the field to assemble roof trusses; guide for truss engineering which is attached showing the engineering loads on each individual truss (beyond the scope of this book)

Sections & Details

shows cabinet elevations; fireplace elevations; elevations of special built-in items like entertainment centers and desks; special trim details; special structural details

Second, a complete bid will contain specifications. These are almost as important as your plans because whereas the plans told you what your new home was going to look like, the specifications will outline the materials that will be used to build your new home and the options your new home will include. A minimal set of specifications will look similar to the set you will find at the back of this chapter that are used in some of our bid proposals to our future clients.

In addition to telling what kind of equipment shall be used and all the items that are included with the price of your home, most categories in a specifications sheet also include allowances – amounts which the client can spend to buy items in that category.

Allowances are the main bid items where builders tend to differ. Because of these varying allowances, the total construction costs from each builder can be considerably different for the exact same home. Also, a few unscrupulous builders have been rumored to use equipment which is cheaper than what is specified so that their total bid amount appears less than their competitors'. This is easy to do when you consider that the minimal appliance package for homes is approximately $500. The appliance package for a custom home can easily cost ten to fifteen times that much, making your appliance package jump to $5000-$7500. And when you discover that industrial appliances can run as much as $10,000 each, mixing and matching equipment can cause a major difference in your total sales price.

Another thing which some builders have been rumored to do is lower the price of allowances so that you have additional costs while the house is being constructed. Several years ago I heard about one local builder who reputedly used this trick to get work from other builders. Although it may have been ethically questionable, it was entirely legal and his clients were powerless to do anything about it because the house would be under construction by the time this came to light. And since they would not be able to illuminate a dog house with the money he provided in their electrical fixture allowance, they would have to shell out hundreds of dollars for that alone. So before they closed on their house, these homebuyers ended up spending more to get the home they wanted than they would have spent initially with the highest bidding builder.

Even when the specifications and the plans are complete, I can still find enough variation in them to be both the low and high bidder in every situation. Because of the endless possibilities that equipment, materials, and allowances provide, the sales price can fluctuate widely. And if you choose the wrong "low bidder," I guarantee you will fight the builder all the way through the job, and your dream home will become the biggest nightmare you have ever encountered. To add insult to injury, once you have finished this monumental battle with the builder, the house will not be anything close to what you wanted, and you will still have to pay for it. And to pay for something you do not want is <u>never</u> a bargain.

Over the years I have given this explanation of why the bid process is faulty to a lot of people, and some of them still have not understood why builders need such detailed plans and specifications. The main reason for this lack of understanding is people do not realize how involved an actual bid on a new home can be. So to demonstrate the complexity of an actual construction estimate, my son and I decided to add one with all material quantities, subcontractor quotes, and labor prices shown on it. This estimate at the end of the chapter demonstrates how many items we have to consider to arrive at the price of a new home when we are given a complete set of plans and specifications. For an actual bidding process, the design engineer or architect would be available to clear up any questions that the bidding builders would have. The architect or design engineer would also be conducting the bid process and oversee all bids when they were delivered. His job would be to review all the bids he received and make sure that all bidders included the same equipment and allowances in their bids, so that he was actually comparing "apples to apples" instead of "apples to oranges".

As you can tell from our estimate[f], there is quite a bit that goes into determining how much a home will cost to build. In addition to material and labor prices, you also have to include miscellaneous items such as insurance, real estate sales commissions, and closing costs. When all of these costs are added up, it should be easy to see why the free-handed scribble we saw at the beginning of the chapter does not possibly have enough information for me to come up with a sales price.

The bidding process works <u>only</u> when the design of the building and the bids are handled by a trained professional. Only an architect or a design professional realizes how much information is needed to come up with an accurate bid and can provide that information to your potential builders. Therefore, if you are not going to use an architect or a design professional, you will not be able to use the bid process effectively.

The use of a registered architect or a design professional throughout the design and construction of your new home is, and will probably always be, the best way to ensure quality. Unfortunately, a registered architect (not to be confused with a draftsman)[1] is fairly expensive; therefore, most homeowners cannot afford one when they build their new home. Instead, most homes in the United States are designed by a builder or a builder working with a local draftsman.

With this in mind, we have written the remainder of our book mainly for homebuyers who will not be using an architect, but rather will be choosing their homes from the stock designs of their builders and making minor modifications to suit their needs. Although homebuyers who use an architect will also benefit from the information we have written, your architect will handle the majority of your new home details for you. Therefore, you will not need as much assistance as the average person who does not have a design professional guiding them through the process.

But regardless of whether you have an architect or not, there is a certain amount of information you as a homeowner <u>must</u> understand before you dive into the designing and home building process. So please refill your glass and read on as I explain what you need to know to avoid a new home nightmare.

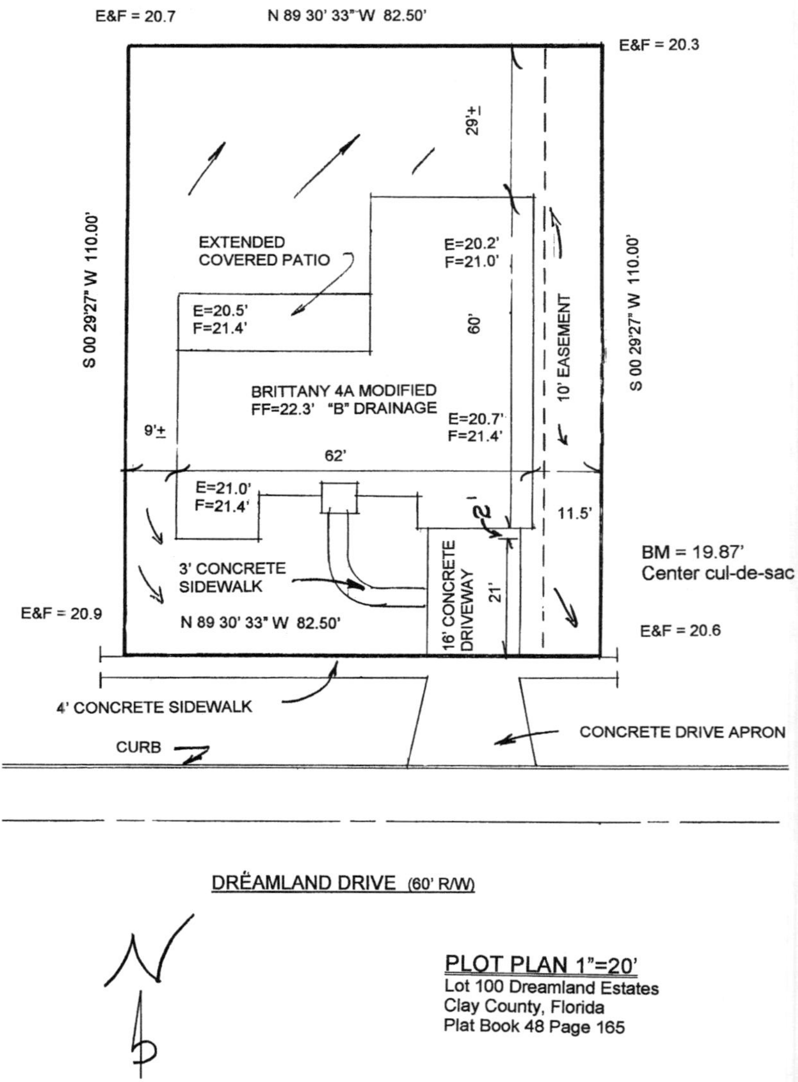

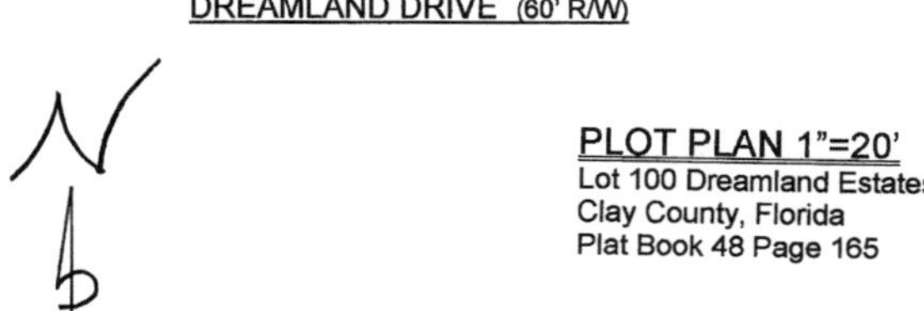

E&F = 20.7
N 89 30' 33" W 82.50'
E&F = 20.3
29'±
S 00 29'27" W 110.00'
EXTENDED COVERED PATIO
E=20.2'
F=21.0'
E=20.5'
F=21.4'
60'
10' EASEMENT
S 00 29'27" W 110.00'
BRITTANY 4A MODIFIED
FF=22.3' "B" DRAINAGE
E=20.7'
F=21.4'
9'±
62'
E=21.0'
F=21.4'
11.5'
2'
BM = 19.87'
Center cul-de-sac
3' CONCRETE SIDEWALK
16' CONCRETE DRIVEWAY
21'
E&F = 20.9
N 89 30' 33" W 82.50'
E&F = 20.6
4' CONCRETE SIDEWALK
CONCRETE DRIVE APRON
CURB
N
DREAMLAND DRIVE (60' R/W)
PLOT PLAN 1"=20'
Lot 100 Dreamland Estates
Clay County, Florida
Plat Book 48 Page 165

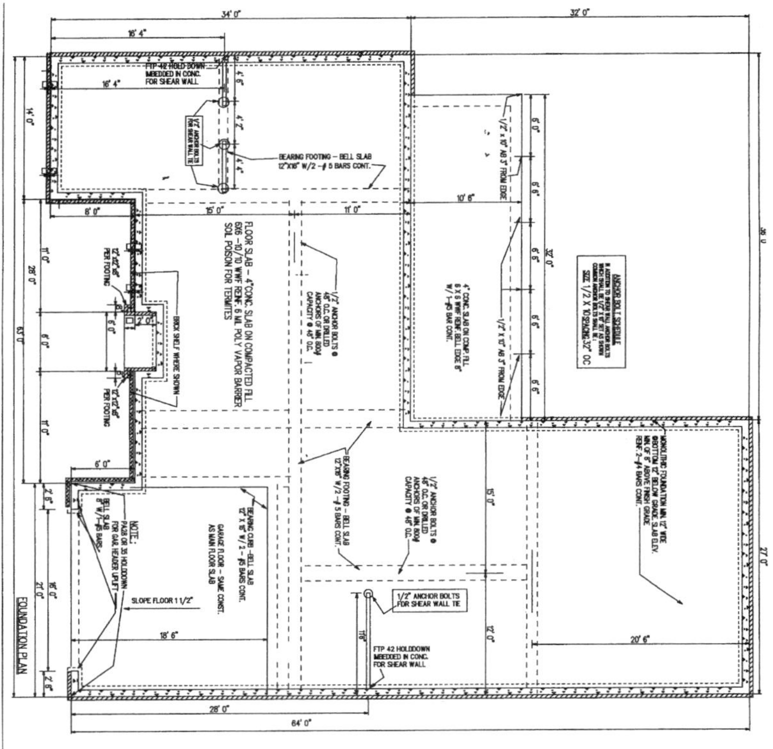

Foundation Plan

Warning: This bid package is for illustrative purposes only. There is absolutely no guarantee that these materials, labor, or other bid item prices and quantities listed herein will build the home plans illustrated in this chapter.

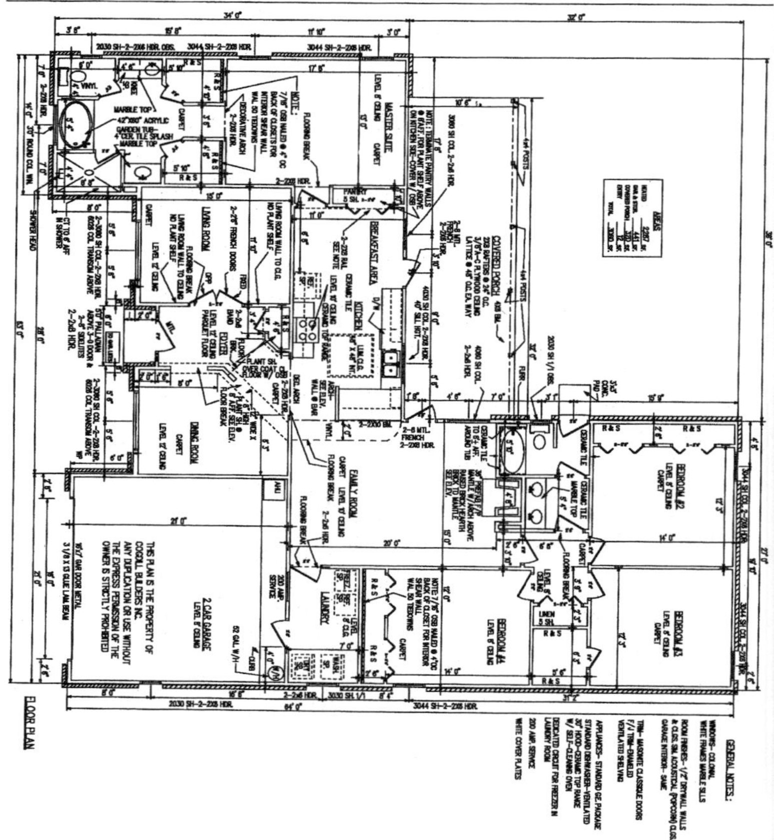

Floor Plan

Warning: This bid package is for illustrative purposes only. There is absolutely no guarantee that these materials, labor, or other bid item prices and quantities listed herein will build the home plans illustrated in this chapter.

FRONT ELEVATION
REAR ELEVATION
SOLDIER COURSE
BRICK VENEER GABLE
ASPHALT SHINGLES
BRICK VENEER
BRICK SILL
DOUBLE ROWLOCK
1'-0"O/H TYP.
ASPHALT SHINGLES
BRICK VENEER
4 X 4 POST
4 X 4 POST
MTL. CHASE COVER
PREFAB MTL. CHIMNEY
7/16" PINE T1-11 SIDING
BRICK SILL
BRICK VENEER
1'-0"O/H TYP.

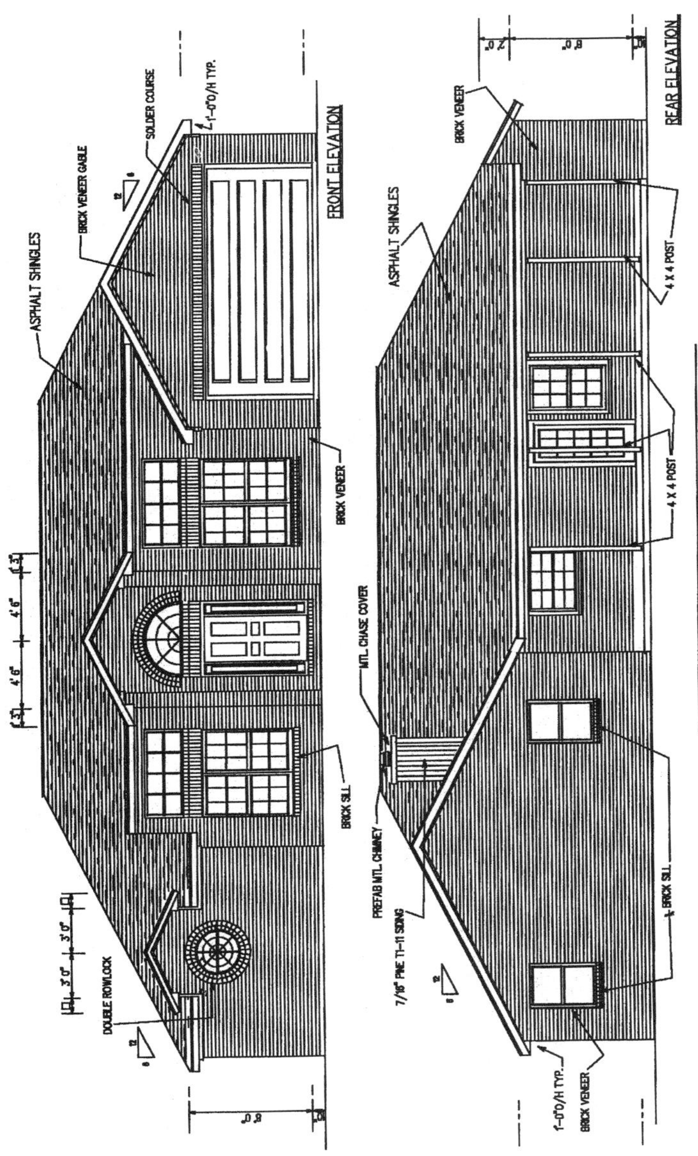

Elevations

ASPHALT SHINGLES
1'-0"O/H TYP.
LEFT ELEVATION
BRICK SILL
BRICK VANEER
CONCRETE PAD
4 X 4 POST
3x3' CONC. PAD
1'-0"O/H TYP.
ASPHALT SHINGLES
RIGHT ELEVATION
BRICK SILL
BRICK VENEER

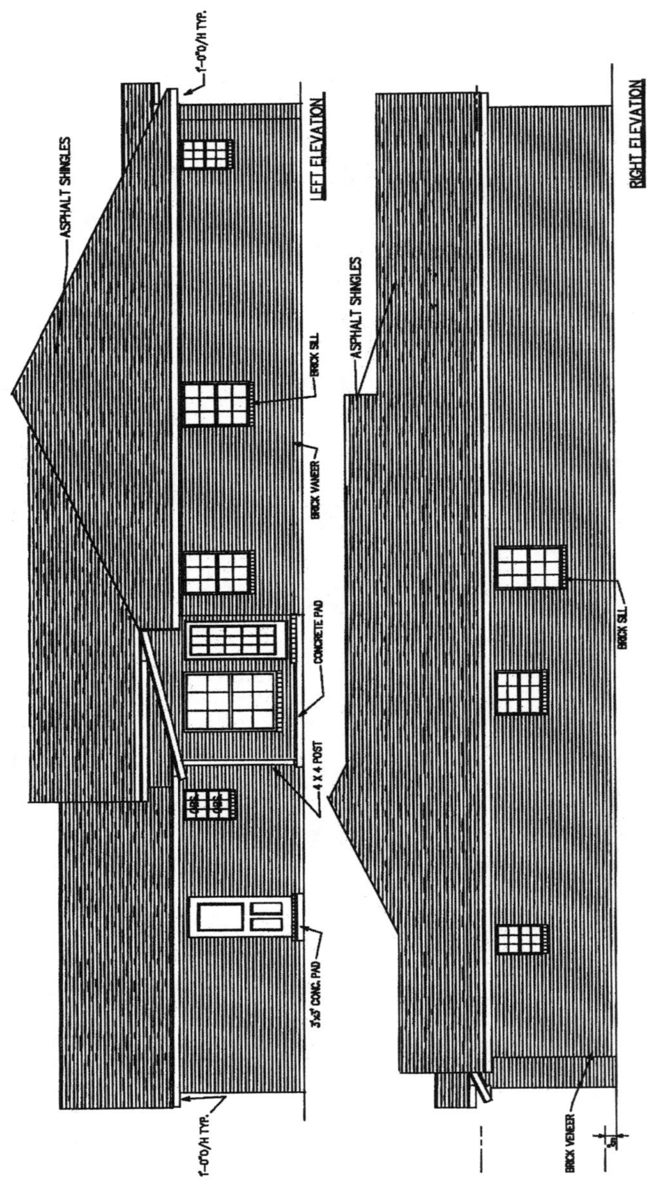

Elevations

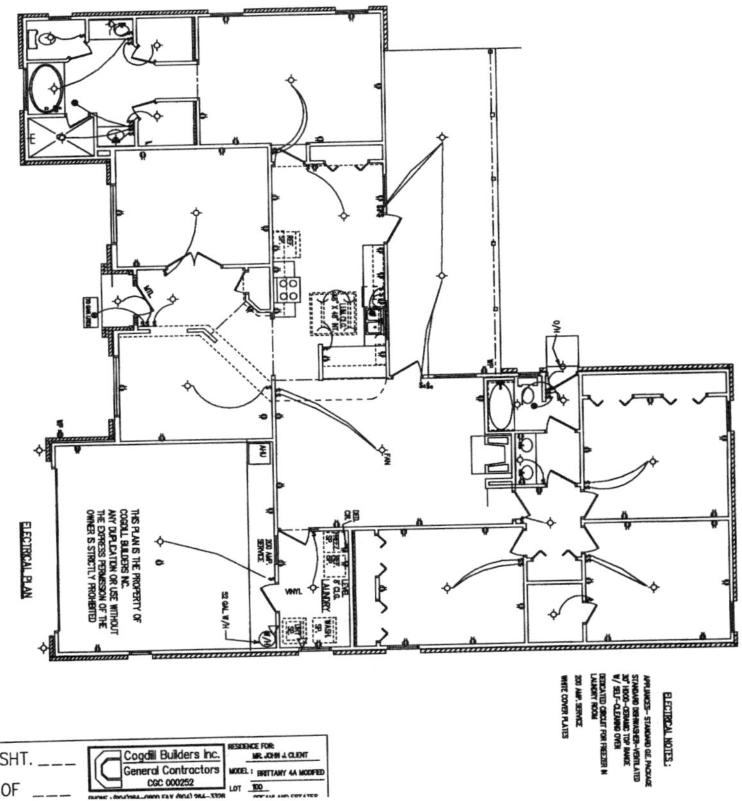

<u>Electrical Plan</u>

Warning: This bid package is for illustrative purposes only. There is absolutely no guarantee that these materials, labor, or other bid item prices and quantities listed herein will build the home plans illustrated in this chapter.

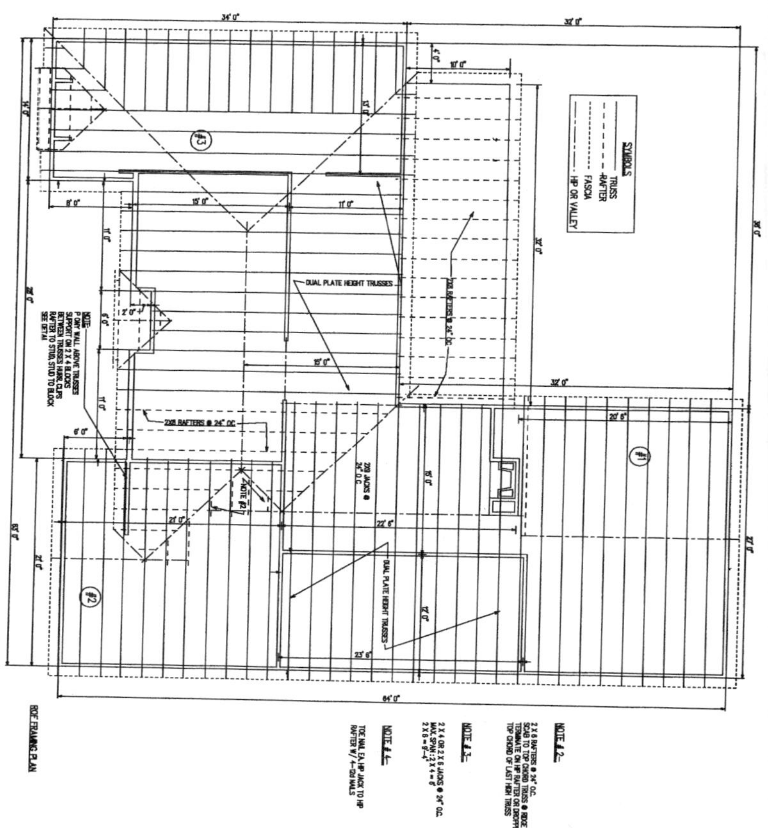

Roof Framing Plan

Warning: This bid package is for illustrative purposes only. There is absolutely no guarantee that these materials, labor, or other bid item prices and quantities listed herein will build the home plans illustrated in this chapter.

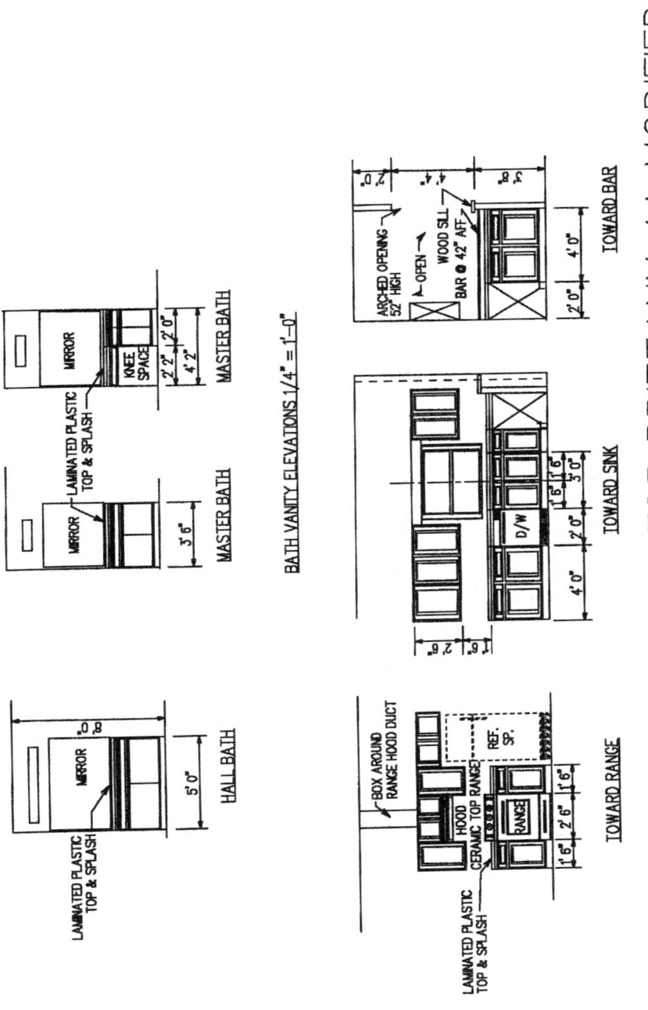
LAMINATED PLASTIC TOP & SPLASH
MIRROR
8' 0"
5' 0"
HALL BATH
MIRROR
LAMINATED PLASTIC TOP & SPLASH
3' 6"
MASTER BATH
MIRROR
KNEE SPACE
2' 2"
2' 0"
4' 2"
MASTER BATH
BATH VANITY ELEVATIONS 1/4" = 1'-0"
BOX AROUND RANGE HOOD DUCT
LAMINATED PLASTIC TOP & SPLASH
HOOD
CERAMIC TOP RANGE
REF. SP.
RANGE
1' 6"
2' 6"
1' 6"
TOWARD RANGE
2' 6"
1' 6"
D/W
4' 0"
2' 0"
6"
6"
3' 0"
TOWARD SINK
ARCHED OPENING 52" HIGH
OPEN
WOOD SILL
BAR @ 42" AFF
2' 0"
4' 4"
3' 8"
2' 0"
4' 0"
TOWARD BAR
CABINET ELEVATIONS FOR BRITTANY 4A MODIFIED
LOT 100 DREAMLAND ESTATES

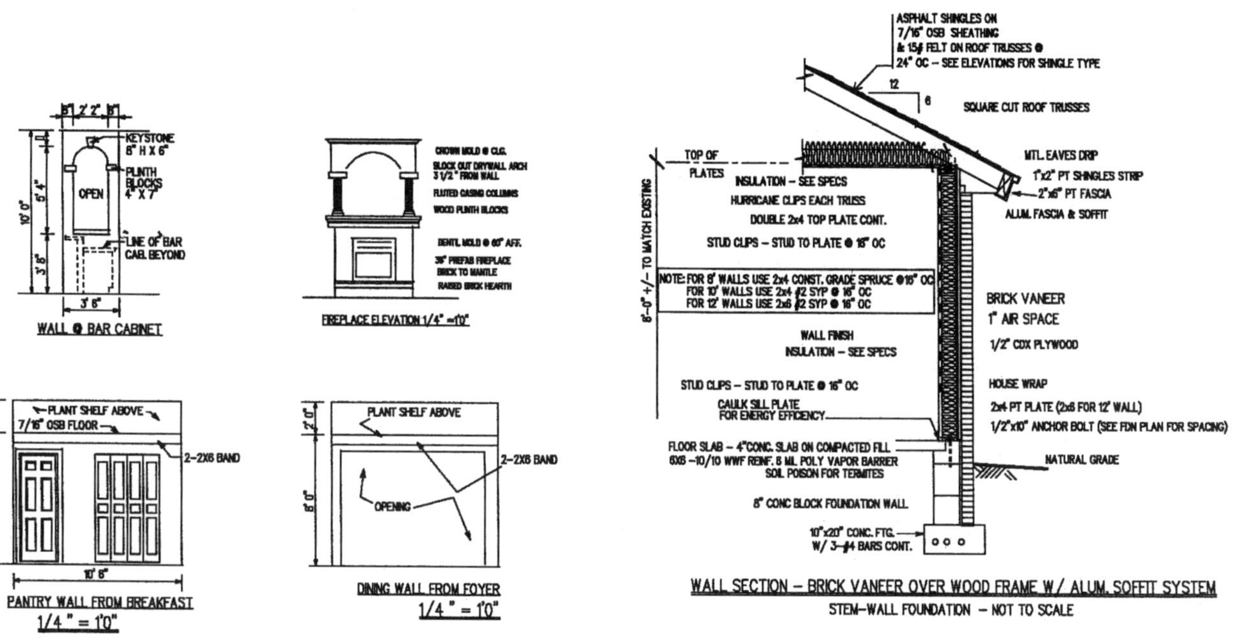
WALL @ BAR CABINET
KEYSTONE 8" H X 6"
PLINTH BLOCKS 4" X 7"
OPEN
LINE OF BAR CAB. BEYOND
10' 0"
5' 4"
3' 6"
3' 6"

FIREPLACE ELEVATION 1/4" =1'0"
CROWN MOLD @ CLG.
BLOCK OUT DRYWALL ARCH 3 1/2" FROM WALL
FLUTED CASING COLUMNS
WOOD PLINTH BLOCKS
DENTL MOLD @ 60" AFF.
36" PREFAB FIREPLACE BRICK TO MANTLE
RAISED BRICK HEARTH

PANTRY WALL FROM BREAKFAST
1/4" = 1'0"
PLANT SHELF ABOVE
7/16" OSB FLOOR
2-2X6 BAND
10' 6"

DINING WALL FROM FOYER
1/4" = 1'0"
PLANT SHELF ABOVE
2-2X6 BAND
OPENING

WALL SECTION – BRICK VANEER OVER WOOD FRAME W/ ALUM. SOFFIT SYSTEM
STEM-WALL FOUNDATION – NOT TO SCALE
ASPHALT SHINGLES ON
7/16" OSB SHEATHING
& 15# FELT ON ROOF TRUSSES @
24" OC – SEE ELEVATIONS FOR SHINGLE TYPE
12
6
SQUARE CUT ROOF TRUSSES
MTL. EAVES DRIP
1"x2" PT SHINGLES STRIP
2"x6" PT FASCIA
ALUM. FASCIA & SOFFIT
TOP OF PLATES
INSULATION – SEE SPECS
HURRICANE CLIPS EACH TRUSS
DOUBLE 2x4 TOP PLATE CONT.
STUD CLIPS – STUD TO PLATE @ 16" OC
NOTE: FOR 8' WALLS USE 2x4 CONST. GRADE SPRUCE @16" OC
FOR 10' WALLS USE 2x4 #2 SYP @ 16" OC
FOR 12' WALLS USE 2x6 #2 SYP @ 16" OC
WALL FINISH
INSULATION – SEE SPECS
STUD CLIPS – STUD TO PLATE @ 16" OC
CAULK SILL PLATE FOR ENERGY EFFICIENCY
8'-0" +/- TO MATCH EXISTING
BRICK VANEER
1" AIR SPACE
1/2" CDX PLYWOOD
HOUSE WRAP
2x4 PT PLATE (2x6 FOR 12' WALL)
1/2"x10" ANCHOR BOLT (SEE FDN PLAN FOR SPACING)
FLOOR SLAB – 4"CONC. SLAB ON COMPACTED FILL
6X6 –10/10 WWF REINF. 8 MIL POLY VAPOR BARRIER
SOIL POISON FOR TERMITES
8" CONC BLOCK FOUNDATION WALL
10"x20" CONC. FTG. W/ 3-#4 BARS CONT.
NATURAL GRADE

SECTIONS AND DETAILS FOR BRITTANY 4A MODIFIED
LOT 100 DREAMLAND ESTATES

Material Estimate

LOT\BLOCK	Lot 100
SUBDIVISION	Dreamland Estates
OWNER	Mr & Mrs John J. Client
27-Dec-99	

ITEM	Quantity	Unit	PRICING Quantity	Unit	MATERIAL Price/Ea	Total	LABOR Price/Ea	Total	SUBCONTRACT Price/Ea	Total
SITEWORK:										
lot clearing	2	ls	2	ls					700	1400
slab fill	648	cy	648	cy					5	3240
slab subgrade level	2	ls	2	ls					150	300
lot fill	180	cy	180	cy					5	900
lot rough grade	1	ls	1	ls					150	150
lot finish grade	1	ls	1	ls					350	350
shrub subcontract	1	ls	1	ls					7000	7000
CONCRETE:										
FOOTING DROP										
#4 rebar	80	ea	80	ea	2.9	232				
3-bar chairs	2	box	2	box	22	44				
tie wire	1	roll	1	roll	5	5				
footing excavation	340	lf	340	lf					2.85	969
footing concrete	24	cy	24	cy	55	1320				
SLAB DROP										
6-mil poly 16' roll	1	ea	1	ea	38	38				
6-mil poly 20' roll	1	ea	1	ea	40	40				
10/10-20/20 wwf	5	roll	5	roll	45	225				
1/2x10 anchor bolts	2	box	2	box	34	68				
#5 rebar	40	ea	40	ea	4.35	174				
slab concrete	55	cy	55	cy	55	3025				
slab finish	2787	sf	2787	sf					0.50	1394
interior footers	170	lf	170	lf					1	170
tractor work	1	ls	1	ls					150	150
driveway grading	2	ls	2	ls					150	300
exterior concrete	40	cy	40	cy	55	2200				
exterior conc. finish	2900	sf	2900	sf					0.5	1450
control joints	1	ls	1	ls					25	25
MASONRY:										
FOUNDATION DROP										
8" header block	250	ea	250	ea	0.95	238			1.2	300
8" stretcher block	1470	ea	1470	ea	0.77	1132			1.2	1764
ashlar block	50	ea	50	ea	0.7	35			1.2	60
4" partition block	500	ea	500	ea	0.7	350			1.2	600
mortar	55	sack	55	sx	6.25	344				
sand	10	cy	10	cy	23.5	235				
EXTERIOR BRICK										
brick	76	cube	38	m	220	8360			400	15200
bay window-solid	2	ea	2	ea					300	600
bay window-half	1	ea	1	ea					150	150
octagonal window	2	ea	2	ea					125	250
Gables-large	6	ea	6	ea					150	900
rowlock	400	lf	400	lf					3	1200
ornamental details	1000	ls	1000	ls						1000
arches	8	ea	8	ea					150	1200
brick column	4	ea	4	ea					180	720
3x3-1/2 angle iron	190	lf	190	lf	2.6	494				
3-1/2x5 angle iron	50	lf	50	lf	4	200				
mortar	390	sacks	390	sx	6.25	2438				
sand	40	cy	40	cy	23.5	940				
wall ties	2	box	2	box	14	28				
FIREPLACE										
42" prefab fireplace	2	ea	2	ea	1000	2000				
brick facing	80	sf	80	sf	0	0			6	480
prefabricated mantle	2	ea	2	ea	65	130	100	200		
mantle brackets	2	pr	2	pr	20	40	25	50		

Material Estimate

LOT\BLOCK	Lot 100
SUBDIVISION	Dreamland Estates
OWNER	Mr & Mrs John J. Client
27-Dec-99	

ITEM	Quantity	Unit	PRICING Quantity	Unit	MATERIAL Price/Ea	Total	LABOR Price/Ea	Total	SUBCONTRACT Price/Ea	Total

ROUGH CARPENTRY:

FRAMING DROP

ITEM	Quantity	Unit	Quantity	Unit	Price/Ea	Total	Price/Ea	Total	Price/Ea	Total
2x4 pt	1000	lf	0.67	mbm	433	289				
2x6 pt	840	lf	0.84	mbm	490	412				
2x8 pt	16	lf	0.02	mbm	800	17				
4x4 pt	170	lf	0.23	mbm	2270	515				
4x6 pt	108	lf	0.22	mbm	900	194				
2x4 spruce stud 10'	450	ea	3.00	mbm	475	1425				
2x4 random utility	2500	lf	1.67	mbm	448	747				
2x4 syp random	1800	lf	1.20	mbm	700	840				
1x4 #2 syp	2000	lf	0.67	mbm	572	381				
2x6x8 #2 syp	9	ea	0.07	mbm	320	23				
2x6x10 #2 syp	370	ea	3.70	mbm	360	1332				
2x6x12 #2 syp	18	ea	0.22	mbm	380	82				
2x6x14 #2 syp	50	ea	0.70	mbm	390	273				
2x6x16 #2 syp	4	ea	0.06	mbm	400	26				
2x6x18 #2 syp	6	ea	0.11	mbm	430	46				
2x8x12 #2 syp	10	ea	0.16	mbm	490	78				
2x8x14 #2 syp	10	ea	0.19	mbm	490	91				
2x8x16 #2 syp	22	ea	0.47	mbm	512	240				
2x8x20 #2 syp	6	ea	0.16	mbm	600	96				
2x10x8 #2 syp	4	ea	0.05	mbm	600	32				
2x10x10 #2 syp	10	ea	0.17	mbm	642	107				
2x10x12 #2 syp	7	ea	0.14	mbm	828	116				
2x10x14 #2 syp	38	ea	0.89	mbm	893	792				
2x10x16 #2 syp	5	ea	0.13	mbm	890	119				
2x10x18 #2 syp	4	ea	0.12	mbm	950	114				
2x12x8 #2 syp	5	ea	0.08	mbm	625	50				
2x12x10 #2 syp	2	ea	0.04	mbm	625	25				
2x12x12 #2 syp	16	ea	0.38	mbm	625	240				
2x12x14 #2 syp	4	ea	0.11	mbm	625	70				
2x12x18 #2 syp	4	ea	0.14	mbm	625	90				
2x12x20 #2 syp	14	ea	0.56	mbm	1200	672				
3-1/8x13-1/2 glue lam	14	lf	14.00	lf	10	140				
3-1/8x15 glue lam	20	lf	20.00	lf	15	300				

ITEM	Quantity	Unit	Quantity	Unit	Price/Ea	Total	Price/Ea	Total	Price/Ea	Total
7/16-4x8 osb	175	sht	5.60	msf	250	1400			3	525
8" beam hanger	150	ea.	150.00	ea	3.04	456				
12d sinker nails	4	ctn	4.00	ctn	21	84				
12d cut nails/lb	15	lb	15.00	lb	1.6	24				
hurricane clips	2	box	200.00	ea	0.18	36				

2nd FLOOR FRAM DROP:

ITEM	Quantity	Unit	Quantity	Unit	Price/Ea	Total	Price/Ea	Total	Price/Ea	Total
2x4 spruce stud	800	ea	4.27	mbm	428	1826				
2x4 random utility	3000	lf	2.00	mbm	448	896				
2x4 syp random	1500	lf	1.00	mbm	522	522				
7/16-4x8 osb	120	sht	3.84	msf	245	941			3	360
3/4x4x8 t&g underlay	15	sht	0.48	msf	700	336			3	45
subfloor adhesive	30	tube	30.00	tube	3.25	98				
20x100 clear visquine	1	roll	1.00	roll	56	56				

TRUSSES:

ITEM	Quantity	Unit	Quantity	Unit	Price/Ea	Total	Price/Ea	Total	Price/Ea	Total
floor truss pkg.	3926	LS	3926	LS	3926	3926				
roof truss pkg.	5867	LS	5867	LS	5867	5867				

ROOF SHEETING DROP:

ITEM	Quantity	Unit	Quantity	Unit	Price/Ea	Total	Price/Ea	Total	Price/Ea	Total
7/16-4x8 osb	180	sht	5.76	msf	245	1411				
felt	35	roll	35.00	roll	10	350			5	175
plyclips	3	box	3.00	box	34	102				
6d commons	1	ctn	1.00	ctn	32	32				
8d commons	3	ctn	3.00	ctn	32	96				
tin tabs	1	ctn	1.00	ctn	39	39				

Material Estimate

LOT\BLOCK	Lot 100
SUBDIVISION	Dreamland Estates
OWNER	Mr & Mrs John J. Client
27-Dec-99	

ITEM	Quantity	Unit	PRICING Quantity	Unit	MATERIAL Price/Ea	Total	LABOR Price/Ea	Total	SUBCONTRACT Price/Ea	Total
CORNACE DROP:										
2x6 spruce	350	lf	0.35	mbm	600	210				
Vinyl Siding Labor	4228	LS	4228	LS						4228
12"x300' moistop	2	roll	2.00	roll	9.25	19				
5x5x10 galv flash	20	ea	20.00	ea	5.25	105				
Framing Subcontract	4800	sf	4800	sf					4.75	22800
ROOFING:										
roofing	6500	LS	6500	LS						6500
4' ridge vent	6	ea	6	ea					75	450
22"x22" skylite	4	ea	4	ea					75	300
special flashing	300	LS	300	LS						300
Gutter	240	lf	240	lf					1.55	372
INSULATION:										
insulation package	3000	LS	3000	LS						3000
TRIM:										
3-0 ext panel door	1	ea	1	ea	1500	1500				
2-8 ext panel door	1	ea	1	ea	185	185				
3-0 simpson 2020	1	ea	1	ea	190	190				
2-8 exterior fr. door	7	ea	7	ea	275	1925				
3-0 int panel door	3	ea	3	ea	42	126				
2-6 int panel door	6	ea	6	ea	36	216				
2-0 int panel door	10	ea	10	ea	33	330				
2-0 panel bifold	2	ea	2	ea	35	70				
4-0 panel bifold	4	ea	4	ea	65	260				
5-0 panel bifold	1	ea	1	ea	75	75				
6-0 panel bifold	1	ea	1	ea	85	85				
Interior panel door	29	ea	29		28	812				
7'int jamb	52	ea	52	ea	11	572				
7' stop	52	ea	52	ea	0.95	49				
7' casing	190	ea	190	ea	2	380				
7' brick moulding	30	ea	30	ea	4.62	139				
3-1/4 base molding	1000	lf	10	clf	60	600				
7/16x11/16 shoe	400	lf	4	clf	19	76				
2-3/4 crown molding	300	lf	3	clf	80	240				
5/8 chair rail:	40	lf	0	clf	120	48				
11/16x14 swp cove	12	ea	2	clf	25	42				
1x4x12 spruce	20	ea	0	mbm	1150	92				
2x6 c-fir r/l	140	lf	0	mbm	2600	364				
2x12 c-fir r\l	60	lf	0	mbm	2600	312				
shim shingles	2	bundl	2	bund	9	18				

ITEM	Quantity	Unit	PRICING Quantity	Unit	MATERIAL Price/Ea	Total	LABOR Price/Ea	Total	SUBCONTRACT Price/Ea	Total
medicine cabinet	6	ea	6	ea	30	180				
16x51 ext blinds	8	ea	8	ea	17.4	139				
entry handleset	1	ea	1	ea	105	105				
grecian level entry	10	ea	10	ea	28	280				
grecian level privacy	14	ea	14	ea	22	308				
grecian level pass.	8	ea	8	ea	20	160				
deadbolt	10	ea	10	ea	20	200				
door bumper	25	ea	25	ea	0.55	14				
4d bright finish nails	2	ctn	2	ctn	30.81	62				
trim labor subcontract	3500	LS	3500	LS						3500
Stair Labor	1500	LS	1500	LS						1500
Wrap Posts	300	LS	300	LS						300
Metal Shelving	360	LF	360	LF					2.5	900

Material Estimate

LOT\BLOCK	Lot 100
SUBDIVISION	Dreamland Estates
OWNER	Mr & Mrs John J. Client
27-Dec-99	

Item	Qty	Unit	Qty	Unit						
GARAGE DOORS										
16X7 metal door	2	ea	2	ea	575	1150				
8x7 metal door	2	ea	2	ea	400	800				
DRYWALL, PLASTER:										
drywall subcontract	12000	LS	12000	LS						12000
TILE:										
tile package	11537	LS	11537	LS						11537
PAINTING & DECORATING										
painting package	5200	LS	5200	LS						5200
WINDOWS/FIXED GLASS										
window package	5882	LS	5882	LS						5882
mirror package	1000	LS	1000	LS						1000
CABINETS										
cabinet package	16800	LS	16800	LS						16800
APPLIANCES										
standard package	3800	LS	3800	LS		3800				
PLUMBING										
plumbing package	7265	LS	7265	LS						7265
acrylic tub	1	ea	1	ea					320	320
marble package	900	LS	900	LS						900
ELECTRICAL										
electrical package	7069	LS	7069	LS						7069
electric fixtures	3000	LS	3000	LS					3000	3000
HVAC										
hvac package	6000	LS	6000	LS						6000
FLOORING										
flooring package	6850	LS	6850	LS						6850
IN-HOUSE COSTS										
drafting time	2	ls	2	ls			200	400		
blueprints	2	ls	2	ls					75	150
extended warranty	2	ls	2	ls					275	550
permitting cost	2	ls	2	ls	0	0			400	800
layout cost	1	ls	1	ls	50	50	200	200		
supervisory time	3	ls	3	ls			300	900		
termite treatment	2	ls	2	ls					550	1100
water meter charge	2	ls	2	ls					220	440
county inspections	3	ls	3	ls					80	240
job cleanup	3	ls	3	ls			360	1080		
dumping fees	3	ls	3	ls			0		300	900
in-house carpentry	10	hr	10	hr			15	150		
MISCELLANEOUS										
security system	3	EA	3	LS					350	1050
central vacuum	5	EA	5	LS					300	1500
CONTINGENCIES	6000	LS	6000	ls						6000

SUMMARY SHEET
LOT\BLOCK   LOT 100
SUBDIVISION   DREAMLAND ESTATES
MODEL   BRITTANY 4A MODIFIED
        27-Dec-99

DIVISION		MATERIAL	LABOR	SUB	DIVISION TOTAL
Site Subtotal		$0.00	$0.00	$13,340.00	$13,340.00
Concrete Subtotal		$7,371.00	$0.00	$4,458.00	$11,829.00
Masonry Subtotal		$16,964.00	$250.00	$24,424.00	$41,638.00
Framing Subtotal		$28,806.00	$0.00	$28,133.00	$56,939.00
Roofing Subtotal		$0.00	$0.00	$7,922.00	$7,922.00
Insulation Subtotal		$0.00	$0.00	$3,000.00	$3,000.00
Trim Subtotal		$10,154.00	$0.00	$6,200.00	$16,354.00
Garage Door Subtot.		$1,950.00	$0.00	$0.00	$1,950.00
Drywall Subtotal		$0.00	$0.00	$12,000.00	$12,000.00
Tile Subtotal		$0.00	$0.00	$11,537.00	$11,537.00
Painting Subtotal		$0.00	$0.00	$5,200.00	$5,200.00
Window Subtotal		$0.00	$0.00	$6,882.00	$6,882.00
Cabinet Subtotal		$0.00	$0.00	$16,800.00	$16,800.00
Appliance Subtotal		$3,800.00	$0.00	$0.00	$3,800.00
Plumbing Subtotal		$0.00	$0.00	$8,485.00	$8,485.00
Electrical Subtotal		$0.00	$0.00	$10,069.00	$10,069.00
HVAC Subtotal		$0.00	$0.00	$6,000.00	$6,000.00
Flooring Subtotal		$0.00	$0.00	$6,850.00	$6,850.00
In-House Subtotal		$50.00	$2,730.00	$4,180.00	$6,960.00
Misc. Subtotal		$0.00	$0.00	$8,550.00	$8,550.00

	MATERIAL	LABOR	SUB	DIVISION TOTAL
Column Totals	$69,095.00	$2,980.00	$184,030.00	$256,105.00
Labor Insurance 30%		$894.00		
Material Taxes 7%	$4,837.00			
Division Totals	$73,932.00	$3,874.00	$184,030.00	

TOTAL CONSTRUCTION
    $261,836

Warning: This bid package is for illustrative purposes only. There is absolutely no guarantee that these materials, labor, or other bid item prices and quantities listed herein will build the home plans illustrated in this chapter.

JOB NUMBER:			
LOT AND BLOCK:	LOT 100		
SUBDIVISION:	DREAMLAND ESTATES	PROFIT/LOSS	$53,302.00
MODEL:	BRITTANY 4A MODIFIED		
BUYER:	MR & MRS JOHN J. CLIENT	PERCENT	
DATE PRINTED:	27-Dec-99	PROFIT:	15.37%
	LOT COST	$35,000.00	
INPUT COSTS:	CONSTRUCTION COST	$261,836.00	
	SALES PRICE	$400,000.00	
	CONSTRUCTION LOAN	$300,000.00	
	CONST LOAN INT.	$4,500.00	
	ORIGINATION FEE	1.00%	$3,000.00
	RECORDING:		
	DEED	$110.00	$110.00
CONSTRUCTION	COPIES	$8.00	$8.00
LOAN	STAMPS	$5.50	$1,650.00
	TITLE INSURANCE	$7.45	$2,235.00
	TOTAL CONSTRUCTION		
	LOAN COSTS:		$7,003.00
	COMMISSION	6.50%	$26,000.00
	LOAN DISCOUNT	1.00%	$4,000.00
PERMANANT	TAX SER/UNDERWRITER	$300.00	$300.00
LOAN	ENDORSEMENTS,DEL.COST	$110.00	$110.00
	TITLE INSURANCE	0.0070	$2,800.00
	RECORDING FEES(FIXED)	$36.00	$36.00
	COPIES (FIXED)	$45.00	$45.00
	APPRAISAL FEE (FIXED)	$0	$0.00
	DEED STAMPS (per $1,000)	$13	$5,068.00
	TOTAL PERMANENT LOAN		
	CLOSING COSTS:		$38,359.00
	CONSTRUCTION LOAN INTEREST		$4,500.00
	CONSTRUCTION LOAN COST		$7,003.00
TOTALS	LOT COST		$35,000.00
	CONSTRUCTION COST		$261,836.00
	PERMANENT LOAN COST		$38,359.00
	CONSTRUCTION TOTAL COST		$346,698.00
PROFIT OR LOSS:			$53,302.00

Warning: This bid package is for illustrative purposes only. There is absolutely no guarantee that these materials, labor, or other bid item prices and quantities listed herein will build the home plans illustrated in this chapter.

<u>Specifications</u>

<u>Residence for Mr. John J. Client</u>
<u>Lot 100 Dreamland Estates</u>

Building permits, utility and mechanical permits are included.

Impact fees[ii], utility hookup fees, assessments, downstream pollution fees, etc. are <u>not</u> included.

<u>No</u> financing or closing costs, construction loan fees or interest are included.

Contractor shall carry Workman's Compensation and General Liability insurance. Builders Risk Insurance shall be carried by the owner.

<u>LANDSCAPING:</u> Allowance for landscaping, sprinkler system (if required), all grass and plants including labor and material, and exterior fill and grading = **<u>Six-thousand ($6000.00) dollars</u>**
Job cleanup will be paid for by builder outside of landscaping allowance.

<u>FOUNDATION:</u> 12"x20" concrete footing with 3-#4 rebars (metal reinforcement bars) continuous.
4" monolithic concrete foundation wall as shown
Standard foundation will be modified to comply with soil engineer's recommendations if necessary and price adjusted accordingly.

<u>FLOOR SLAB:</u> 4" concrete slab on compacted fill[iii]– bearing footings 12"x20"
2-#5 rebar continuous – 6x6 – 10/10 WWF reinforcement
Soil poisoning by XYZ Pest Control
$100,000 termite bond provided by pest control company

**EXTERIOR
CONCRETE:**   4" concrete slab on grade – allowance of 2,000 square feet (SF)

**EXTERIOR
WALLS:**   Brick veneer over wood frame as shown in wall section

ROOF: 30-year asphalt shingles

GUTTERS: Aluminum continuous gutters at all openings and across back porch and garage

**INTERIOR WALLS
& CEILINGS:**   ½" drywall taped and finished – knock down ceilings throughout

INSULATION: R-11 exterior walls, R-30 ceilings

**INTERIOR
FINISH:**   Latex paint over drywall
Semigloss enamel on all interior trim
Wall covering (wallpaper) not included

WINDOWS: Aluminum tilt-out windows, double glazed as manufactured by XYZ Aluminum – all colonial muntins[iv]

**EXTERIOR DOOR
& SIDELITES:**   **Twelve-hundred ($1,200) dollar allowance[v]**
Metal garage doors as shown
For 1/3 horsepower opener **add $515.00**

**INTERIOR
DOORS:**   Classique or Colonist style panel doors
Williamsburg style trim – finger jointed

CABINETS: Raised panel as shown
Marble vanity bath tops
Laminated plastic kitchen tops.
Allowance for cabinets (bath & kitchen, not including tops)
Three-thousand six-hundred ($3,600) dollars

FINISH FLOOR: Allowance of **Five-thousand ($5,000) dollars** for carpet & vinyl flooring
Ceramic tile flooring baths and foyer

CERAMIC TILE: For baths and shower walls (material only)
For bath floors = $2.00 per sq. ft.
For Foyer floors = $2.50 per sq. ft.

PLUMBING: Marble vanity tops with integral sinks (4 total).
Marble tub master bath with deck mounted trim
Single lever Delta® brand faucets in vanities, showers, hall bath tub
Elongated water closets
Fiberglass laundry sink
Americast® brand porcelain kitchen sink
52 gallon high recovery water heater
Tub enclosure hall bath
Shower doors and enclosure as shown

ELECTRICAL: 200 amp service
Prewire and hang 7 fans (cost of fans in fixture allowance)
Wiring as shown including freezer and ironing board circuit
Phones, TV outlets, security system by others
Allowance for light fixtures including fans =
Twenty-five hundred ($2,500.00) dollars

<u>HEATING & A/C:</u> Lennox air-to-air heat pump – 4 ton – SEER 12.[vi]
For SEER 13 **<u>add four hundred ($400.00) dollars</u>**
(Standard builder units are SEER 10)

<u>APPLIANCES:</u> **<u>Allowance = two-thousand ($2,000.00) dollars(with tax)</u>**

f **Warning: This bid package is for illustrative purposes only. There is absolutely no guarantee that these materials, labor, or other bid item prices and quantities listed herein will build the home plans illustrated in this chapter.**

Highlights for Chapter 2

- The bid process almost never works correctly because most people do not realize how detailed a bid must be and cannot properly evaluate bids.

- A complete bid package must include a full set of plans and specifications that cover all materials and methods used to construct your home.

- The allowances your builders give for appliances, finish flooring, and so forth should be stated and the same amounts included in each bid.

- To use the bid process effectively, you must have a designer/architect who designed your home available to answer any questions the builders have who are estimating and bidding your home.

- Allowances are the main items where builders tend to differ in their bids.

- Because of the endless possibilities that equipment, material, and allowances provide, the sales price of your new home can vary widely.

- The bid process works only when the design of your home and the bid process are handled by a trained professional.

- If you are not going to use an architect or design professional, then you **will not** be able to use the bid process effectively.

- Most homes in the United States are not designed by a registered architect. Instead, most are designed by a builder or a builder working with a local draftsman.

- This book is written primarily for those who will not have a registered architect handling the majority of their new home details for them. Although homebuyers using an architect will also benefit from what we have written, they will not need as much assistance as those who do not have a design professional guiding them through the process.

i   A registered architect is licensed by the state in which he practices, usually holds a college degree in architectural design and must go through an apprenticeship program prior to taking the state board exams.

A draftsman is not required to have a professional license or a college degree and may have varying amounts of experience – thus it is harder to judge the competence of a draftsman.

ii   Impact fees – fees imposed by some local municipalities on any type of new developing or building to cover the cost or "impact" such building will have on the local community and its resources (water, sewer, fire department, etc). Such fees are highly controversial and may become an increasing source of new home price increases in the near future.

iii   Compacted fill – the soil underneath the slab is brought to the jobsite and spread by machines in layers 8" thick until the proper slab height is reached. The weight of the machines going over the dirt as well as the dirt being crushed by a vibrating machine called a "tamper" compacts the dirt to help prevent the slab from cracking and the home from settling due to voids in the dirt.

iv   Muntins – pieces of aluminum placed between the double panes of glass to make the window look like it is made up of several smaller pieces of glass instead of one solid piece (makes the window look more traditional).

v   Allowances- a specific amount of money set aside in a construction contract for certain categories of construction items (i.e. lighting fixtures, appliances, landscaping, etc)

vi   SEER – Summer Energy Efficiency Rating – The higher the SEER rating, the more efficient your air conditioning unit. However, the higher initial cost of purchasing a unit with a higher SEER rating may offset the energy savings the unit will provide. Ask a local heating and air-conditioning contractor for advice before choosing an A/C unit for your home.

Notes

Chapter 3:
So Now What?

Now that we have shot down many of your presumptions about designing and building a new home, the next question you have to ask yourself is "So Now What?"

Do I really want to go through with the long, laborious process of creating a new home?

Because the truth is that no matter how much you know or think you know about it, you are going to face a lot of frustration making your home come out exactly the way you envision it. And as you read through this book and realize that the process is much more involved than you previously thought, the next question that pops into your head is, "Why should I do it?"

Well, the answer to these questions is, Yes, you should do it and it is worth doing for a number of reasons. First, designing and building your own home is the only way to get exactly what you want. No other person on the planet has the exact same needs and desires that you do. Therefore, no one else will put the same amenities and extras you desire into their custom home. So if you want your home to include everything you are looking for, you will not find it in a house that was built for someone else. You will only find it in a home you create for yourself.

Second, tailoring your home to fit your specific needs is very important to you. How do I know this? Because if it did not matter what kind of house you lived in, you would take the first one that came along which could accommodate your family and all of your belongings. Also, you would not have read this far into a book about designing and building a custom home.

Finally, when all goes right with a new home, you get a brand new product with new warranties. New home warranties from your builder and component manufacturers allow you the peace of mind that comes with knowing you are protected if something should go wrong in your new home.

With a used house, you get an old product with no warranties and a questionable remaining life. And when something breaks down you will be completely on your own. The only person with peace of mind in a used house is the repairman, because he knows he will make quite a bit of money when something eventually and inevitably breaks down.

So when you consider that a new custom home will give you exactly what you want, with new materials and warranties, you will find that designing and building a new home is well worth the extra effort. Your home is something that will last for many years and provide the background for your fondest moments with friends and family. Such special memories require special surroundings, and what place could be more special than the home you create from the ground up?

When you rely on your good judgement, carefully research and choose your builder, and get answers to all your questions, you will discover that building and designing your new home is a lot of work that will pay off in the long run. So please do not get discouraged. Just relax and get comfortable as we begin to learn how to make designing and building your new home a dream instead of a nightmare.

__Highlights of Chapter 3__

- No matter how much you know or think you know, you will face a lot of frustration and aggravation making your home exactly the way you want it.

- Designing and building your custom home is the only way to get the home you want.

- With a new home you get a brand new product with new warranties.

- With an used house you get an old product with no warranties and a questionable remaining life.

- When you rely on your good judgement, carefully research and choose your builder, and get answers to all your questions, you will discover that designing and building your new home is a lot of work that will pay off in the long run.

Notes

Notes

Chapter 4:
Setting Your Budget

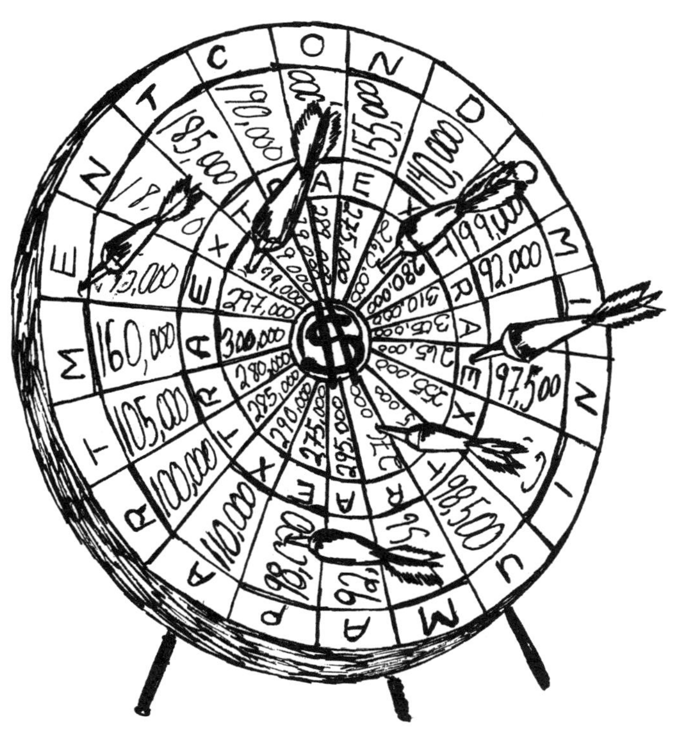

I remember a couple I met a few years ago who were supposedly ready to choose their lot and start designing and building their new home. After about four hours of deciding what model they wanted to customize, we signed a contract and I started to head back to my office. The wife then tapped me on the shoulder and said, "By the way, I think the mortgage company made a mistake."

"Why do you say that?" I asked.

"Because they said that our credit report showed several past due bills that were hurting our credit. But that can't be right because the bankruptcy we had two months ago took care of them."

Thinking back on this now it seems really funny, but at the time I had a much different reaction. Those several hours I had spent working with them were down the drain simply because these people would not be able to obtain a mortgage. They knew exactly what they wanted and had determined everything they needed for their new home – they just did not have a way of paying for it.

When you begin planning your new home, the first thing you need to do is ask yourself, "How much home do I feel comfortable building on my budget?" Most of you reading this should have an idea what the answer is because you are most likely paying rent or paying on a current mortgage. Also, looking at new homes in the area where you intend to build will give you a feel for how much homes cost and how much home you can reasonably expect for your money.

Once you have developed a feel for your budget, you should call two or three mortgage companies and talk to a loan officer. They will give you a good idea as to what kind of mortgages are available and for how much you can qualify. Traditionally, the mortgage company has been involved in the home buying process only after the contract between the homebuyer and the builder is signed. In the last few years, however, the trend has been to involve the mortgage company in the process before the homebuyer even begins to negotiate with a builder. This practice of making a loan application for a home mortgage before you begin to search for a new home is called prequalifying.

The prequalifying process works very well because it lets you know <u>exactly</u> what you can qualify for and what your monthly mortgage payments should be. It also tells you what your closing costs will be and allows you to get the application for your home mortgage out of the way <u>before</u> you begin to look for a home – saving you headaches and letting you know exactly how much money you have to "play with".

Because mortgage programs and rates offered to homebuyers change on a monthly or sometimes even a daily basis, you will need to call a mortgage loan officer so they can help you get the right program for your needs. The numbers and types of mortgage programs available to homebuyers today are greater than ever and would fill many books this size. So instead of giving you a long list of mortgage information that would make a technical manual seem interesting, I want to give you a short list of things you need to do to prequalify for your home mortgage.

To get the ball rolling, you simply need to call several mortgage companies and make an appointment with a mortgage loan officer. Take along the information you need to make a mortgage application and let the officer fill out the application for you. Although the information you need to take with you to a mortgage application may vary from company to company, please refer to our list of typical items at the end of this chapter to get an idea of what your lending officer will need when you make an application for your mortgage.

Once you have made your application and completed the prequalifying process, the only thing you will need to finalize your mortgage financing is a copy of the Purchase & Sale Agreement (contract) on your new home. Your mortgage loan officer will add this agreement to your prequalifying paperwork and get your mortgage finalized so you can start building your new home.

During your prequalifying meeting, your mortgage officer will discuss several financing methods for your new mortgage. While you are listening to your options, you need to keep an open mind and listen to everything you are being told very carefully. This way you can avoid inadvertently paying more per month because of a simple misunderstanding.

Several years ago when interest rates were at a peak, a number of "creative financing" methods were invented to make new home mortgages easier to obtain. Most of these creative financing methods were based on the idea of forestalling higher premiums so that homebuyers could qualify under an initially lower-than-market interest rate that would steadily increase to a higher-than-market interest rate later in the mortgage. Mortgages of this type were called Adjustable Rate Mortgages or ARMs and caused a lot of problems for homebuyers because they did not have a maximum limit on how high the interest rate could be raised. This led to people paying an artificially low interest rate when they moved into their home and then paying an exorbitant interest rate after they had lived there for three to five years. Some of these people did not fully understand how their ARM worked and discovered the hard way that they should have paid more attention to their mortgage lender.

Today's ARMs have a maximum limit on how high your interest rate can climb and thereby avoid the problem of exorbitant interest later in the mortgage. The Adjustable Rate Mortgages of today are therefore not something to be looked upon with suspicion, but rather can be desirable in some circumstances.

When you talk to your mortgage lender, keep an open mind as to what kind of financing options will work best for you. Although most people want to stick with a conventional fixed rate mortgage, you may find an alternative method such as an ARM that better fits your needs. But whether you use a conventional mortgage or an alternative method, fully discuss your options with your mortgage lender and make sure you understand <u>exactly</u> what kind of terms and payment plans each mortgage has to offer. If you do not understand something, ask your lender to explain it to you. A lender would much rather spend some extra time with you before you make a mortgage application than allow a misunderstanding to fester into a bigger problem that could easily have been avoided.

Because mortgage lenders often speak in tongues when you are not familiar with the lingo, we would like to give you some basic terms that you may hear during your prequalifying meeting with your mortgage officer. You will find this Glossary of Mortgage Terms in the Appendix of this book where it can be easily referenced when it comes time to prequalify for your mortgage.

Those of you who own your home or have most of your mortgage paid off need to stop at this point and evaluate your situation closely. Call a local real estate professional and have them do a free market study to determine the approximate value of your home. In this study they will compare your home with other similar homes in the area that have recently sold and come up with a reasonable selling price for your home. Armed with this information and knowing how much money you have to pay on your current mortgage, a real estate professional will be able to calculate how much money you can expect to receive once your home is sold.

When your real estate professional does an evaluation of your existing home, please keep in mind that he may come up with a selling price that is less than what you think your home is worth. Often homeowners have deep sentimental attachments and place more value on their home's special features than a prospective buyer who is seeing their home for the first time. Oftentimes I have heard people say, "We just paid thousands of dollars to have new carpeting installed throughout our home. It was $50 per yard, the best quality carpet available. That should make our home more valuable, right?" It should, but the reality is that it will not. A stranger coming into your home and seeing your carpet for the first time does not realize or may not care that your carpet is the best quality available. To most people, carpet is simply carpet. So even though you may have paid top dollar, chances are you will not receive top dollar.

This is why you need to keep an open mind when your real estate professional evaluates your home. A lot of expensive amenities such as swimming pools and wood decks may not be well received by the buying public. Keep this in mind when you are setting a sales price for your home so that the price you ask is reasonable. A good real estate professional will help you set a realistic sales price and will tell you what you can expect to receive for your home, even if it is not the price you want to hear.

Oftentimes it is very important to check with two or three real estate professionals to make sure that the estimated sales price they give you is a realistic price for your area. Remember that you can ask as much for your home as you want, but if you do not find a buyer willing to pay that price, you are completely out of luck.

Several years ago a rumor circulated about a local real estate "professional" who took advantage of home sellers by telling them he could sell their homes for much more than he actually could. For example, a typical home seller who was going to be transferred out of town and needed to sell his home fast would call two or three real estate professionals for help. These professionals would conduct careful market analyses and tell the seller they could get maybe $70,000 to $75,000 for his home. Then the seller would call this "professional" and he would tell them, "Don't worry, I can get you anywhere from $90,000 to $95,000 for your home easy. Those others you called did not know what they were talking about. Just list your home with me, and I should have it sold in two or three months tops."

Six months later, the seller had moved his family to a new town and was getting crushed under the financial weight of making two separate mortgage payments. By this time he was desperate to sell his home, and the "professional" would come to him with an offer for $65,000.

"But I thought you said you could get me at least $90,000," the seller would exclaim.

"Well, I thought I could too but the market has just gotten a little soft here since your home went up for sale. Unfortunately, this is the best I can do. Do you want to take it or keep making two mortgage payments for the next six months?"

"Just take it and have done with it," the seller would say dejectedly.

Things like this are completely unethical but unfortunately are not illegal. The "professional" can just say that he thought he could get that much for the home and **OOPS!** – he was wrong. But he is really sorry. So instead of being in a sorry state when you cannot sell your home for what a real estate professional quoted, ask several professionals in your area and see if the prices they quote agree with one another. Because if one of them tells you a price that seems too good to be true, it will be - 99 times out of 100.

Once you know how much you can reasonably receive for your home, you may have several options open to you. The home you live in now may be nicer than the new homes you can afford in your area. In that case, you may find that moving to a new home may not be worth it. Conversely, you could also find that the value of your current home allows you to expand your budget and buy a nicer home than you originally planned. This "added bonus" of home ownership could make your day when you find you can qualify for a larger mortgage than you thought.

A few visits to model homes and communities in your area will give you a cursory idea about the selling prices of new homes. Even though you are going to customize your new home and change the sales price from those homes you view initially, you will get a feel for how much you need to budget.

It is extremely important that the budget you set for your new home be realistic. You must know how much you can qualify for, the monthly payments you can handle, the down payment you can come up with, and how well you can afford the routine monthly expenses for your new home. Because if you are not realistic in your approach to budgeting for your new home, the rest of the designing and building process will come to a screeching halt.

<u>Information Needed to Make An Application for Mortgage[i]</u>

For a complete list of mortgage terms, look in the Appendix section under Glossary of Mortgage Terms.

- W-2 forms for the past two years
- Most recent 30-day paystubs
- Payment history for past two years to landlord/mortgage company
- Employment information for the past two years (names and addresses)
- Bank name, account numbers, and balances of all bank accounts
- Names, addresses, account numbers, balances and monthly payments of all open loans
- Addresses and loan information on other real estate owned
- A check to pay for a credit report and appraisal
- Bank statements for previous 3 months on all bank accounts
- Complete divorce decree, if necessary
- Your Purchase and Sale Agreement (Contract) and legal description of the home you are buying (if prequalifying, these can be added to your loan package once you have found your new home and your mortgage application completed at that time)
- Copy of Driver's License or other photo identification (for FHA/VA loan)
- Social Security Card (FHA Loan only)
- Certificate of Eligibility or DD214's (VA Loan only)
- Statement of Service (VA only)
- If commissioned or self-employed, last two years tax returns with schedules, year-to-date P&L and Balance Sheets

Highlights of Chapter 4

- When you begin planning your new home, ask yourself, "How much home do I feel comfortable building on my budget?"

- You should have a good idea based on your current mortgage or your current rent payments.

- After getting a feel for your budget, call two or three mortgage companies and find how much you can qualify for and what types of mortgages are available.

- Prequalifying or making a loan application before negotiating with a prospective builder, is good to do because it lets you know how much you qualify for <u>before</u> you look for a new home.

- To prequalify, call several mortgage companies and make an appointment with a loan officer. Take along the information you will need to make a mortgage application (see the list at the end of this chapter) and let the officer fill out the application for you.

- During your prequalifying meeting, be sure to listen to all the options your mortgage loan officer explains to you very carefully. Although most people go with a conventional fixed rate mortgage, you may find an alternative method that better suits your needs.

- For handy reference, please refer to the Glossary of Mortgage Terms in the Appendix of this book for easy definitions of mortgage terms.

- Those who own their own home need to call two or three real estate professionals and have them do a free market analysis to determine how much your home is worth.

- When this analysis is done, the professional may come up with a selling price less than you think your home is worth. This is because a lot of expensive amenities such as swimming pools and wood decks may not be well received by the buying public.

- It is important to call two or three real estate professionals and compare the estimated selling prices for your home. If one of them is much higher than the others, chances are that price is too good to be true.

- Once you know what you can reasonably receive for your current home, you may find your home is nicer than new homes in the area and you want to stay where you are. Conversely, you may find your home's value allows you to buy a nicer home than you originally planned.

- It is extremely important that the budget you set for your new home be realistic. Otherwise, the designing and building process will come to a screeching halt.

[i] Charter One Mortgage, A Subsidiary of Charter One Bank, F.S.B., 6622 Southpoint Drive South, Suite 180, Jacksonville, FL 32216.

Notes

Notes

Chapter 5:
Choosing An Area

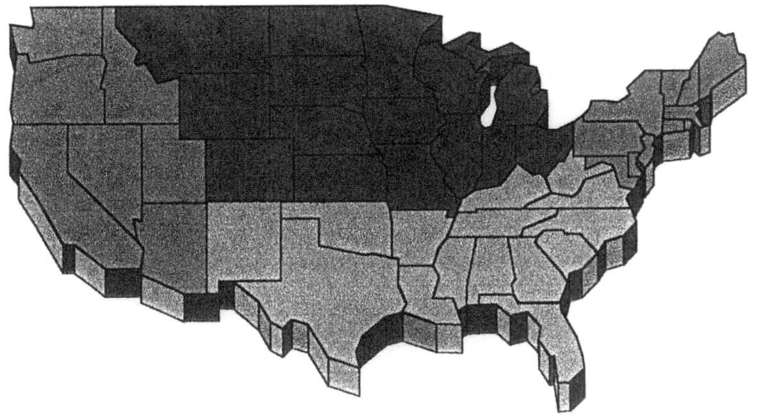

The importance of researching an area where you want to live can never be overemphasized. It affects everything from how comfortable you feel when you walk out your front door to whether you can get your groceries home before the ice cream melts and the milk goes sour. With such an important investment as your new home, you should consider several things carefully before you select the area of town your family will come home to every evening.

➤ Where you and your spouse work

One of the surest ways to tell that you have moved to the wrong area is having to wake up at 5 A.M. to get to work after getting home at 9 P.M. the night before. Several months of this grueling schedule will be enough to tell you that moving to this area was not one of your better ideas.

To avoid this, the best thing to do would be to keep in mind how far you will have to drive to work when you decide where to build. Traditionally, job location has been the main factor in determining the area where you want to live. As new housing suburbs have gotten further and further away from town, the morning and evening commutes of working people have gotten longer and longer. Therefore, if you choose to move to a new suburban community, you can expect the length of your commute to be slightly longer than your current drive.

Oddly enough, the length of the commute is not as important as the way you personally feel about your commute's length. For example, I know several people who drive over an hour to get to work every morning. These people consider the extra time on the road as a relaxing way to begin and plan the day in the morning and a luxurious way to unwind in solitude on the way home.

On the other hand, we have lost home sales to people who claimed that a twenty or thirty minute commute was too long. So it is more a question of your individual attitude toward the driving distance than how much time you have to actually spend on the road.

Your attitude toward the commute may also be affected by the infrastructure of your area. For example, you may not mind an extra twenty minutes added to your drive if it consists of well-designed interstate highways or lightly traveled back roads. But when your new route adds an extra five minutes of bumper-to-bumper traffic to your commute, you may find it to be the longest five minutes of your day.

To judge the severity of your new commute, you and your spouse may want to drive to work one morning and one afternoon during rush hour to see just how much of a drive you may be facing. Equipped with a map of the area, you may find some alternate routes you can take that may be a longer distance but allow a shorter driving time. Or you may realize that although the commute is more of a distance than you would like, the area offers other amenities that are more important to you and your family.

➤ Proximity and Quality of Schools and Daycare

Sending your children to a good school is one of the most important ways to ensure that they will have a successful future. So naturally the next thing you should look for is the quality of the schools your children will attend. Kindergarten, elementary, junior high, high school, and local colleges – all of these levels should be checked to ensure that your children will have a great education. Also, the ratings of the local private as well as public schools should be investigated so that your child has several quality options when the time comes for them to pursue their education.

Even for "empty nester" couples whose children are grown and couples who have no children, the quality of area schools is important for one simple reason. When the quality of local schools is poor, the resale value of area homes suffers. This is because families with children who look at your home in the future will want their kids to attend good schools. And if this not possible living in your home, they will keep searching until they find a home where it is possible.

If your family relies on two incomes or you are a single parent with a young child, the quality of local daycare centers will be a key issue. While you are in the area visiting and researching local schools, it would be wise to take the time and visit local daycare centers as well. As many child care professionals will tell you, the early years of a child's life are the most important in their growth and development. So careful investigation of the people who are going to be influencing your children during their formative years must be a priority.

➤ Recreation/Activities

One of the biggest things you need to think about when choosing an area of town is what kind of leisure activities you like to do and their availability in your new area. Sometimes the proximity to a boat ramp might be enough to pull you from one area to another if you like to spend time on the water. Or you may find a local theater group that captures your heart as well as your entertainment dollars. And, of course, for families with young children, supervised activities such as Little League teams or monitored playgrounds provide an opportunity for your child to interact with other local children in a safe, fun environment.

To find all the amenities a new area has to offer, perform small investigative tasks such as checking with the local Chamber of Commerce, local websites, or even just drive around for a day. Activities such as these will help you and your family get acquainted with your new area and make moments together more enjoyable for all of you.

➤ Distance from local shopping

Traditionally, only women have considered how far they have to drive to shop, but nowadays it is important for men as well. With both men and women working to make ends meet, it is important to each of them to have nearby shopping so they do not have to drive an hour to buy a gallon of milk.

Long drives in the country may be fine for lazy Sunday afternoons, but not on a Wednesday night at 7 P.M when you have just gotten home and realized that you are out of dog food. Then a long drive can become less of a leisure activity and more of a laborious chore that makes you think twice about where you have chosen to call home.

➤ Churches & Friends

For many people, church is one of the most important parts of their lives. In addition to being where you go to strengthen your beliefs and reaffirm your reasons for living, it is also where most friendships are formed. Therefore, when you move to a new home, you need to keep in mind how far you will be from church. Since you may be making this trip at least once a week, you will need to keep this routine journey manageable.

➢ Resale Value

Although you may have no intention of ever selling your home, circumstances may arise which force you to change your mind. For example, you may get a promotion that requires you to move out of town or inherit a mansion from your long lost uncle and be forced to sell your home to pay the taxes. Besides, chances are high that your home will be resold down the road because even if you do not sell it yourself, the house will probably be sold off with the rest of your estate after your death. So a little preparation now can pay big dividends down the road and allow you to use your home as a financial "stepping stone" if you decide later to sell it and move into something different.

One of the best ways to ensure that the resale value of your home will continue to rise is to carefully choose the place to construct it. A good resale area will be a growing area where property values are increasing and new and proposed buildings will help raise the property values even higher. If you already live in the city where you are planning to build, it is probably obvious to you where the high growth areas of the city are and whether the growth in these areas is increasing the property values for local landowners. As long as you are in an upscale, growing area, you can rest assured that the value of your home will most likely continue to rise long after you and your family have moved into it.

If you are not familiar with the area where you are planning to move, it would probably be wise to ask a local real estate professional for help in finding the upscale, growing areas where you should build. Pointing out which areas of the city are increasing in value is one of the most important functions a real estate professional can do for you. A knowledgeable professional can often save many hours of legwork and help you avoid costly mistakes.

Another source of valuable local information would be the local Chamber of Commerce and the local tax assessor's office. These agencies can tell you such things as average property values, average incomes in the area, local demographics, and how all these factors have been changing over the last several years. This information will tell you how the area has been developing and should give you a good idea how the area will develop after you have settled there.

➤ Trade-offs

When you begin searching for the perfect location, you are going to find that like all things perfect, it does not exist. In order to get the best location possible, you are going to have to sit down and decide which of the items we discussed are the most important and look for a location with those characteristics. For example, you may find that to live in a better quality neighborhood, you have to allow yourself some extra driving time into work. Or you may have less driving time to get to work, but you may have to live farther from your favorite fishing hole.

Before you begin to look for your new home, sit down with your family and make a list of things that are important to you. Then prioritize these items and keep this list in mind as you search for your new area. If you realize from the outset what is important to you and keep it in mind when you find a place to build, you will enjoy your new home much more.

➤ Other Considerations

When you have discovered an area of town where you would like to live, it is important that you drive through the various local communities and discover how many empty homesites are available. You must remember that when you have found the perfect area of town, you have to also find an empty plot of land to build on or your search will have been in vain.

In addition to the availability of your homesite, you also must find what the sales price of homes in your new area will be and if they are within your new home budget. For example, when you check out new home communities, you may discover that new home prices start at $300,000 and up. This is fine if your budget allows you to buy a $400,000 home, but if you can only afford a $200,000 home, then you will need to look for a less expensive area.

Such careful and time-consuming research on a new area may seem excessive or maybe even a little frustrating at times. And after a long weekend of driving around different communities and visiting countless builder models, you may feel the urge to just walk into the next house that halfway fits your needs and sign a contract. During these times you must remember that after living in your new home for several care-free years, the time you take now will seem minute compared to the peace of mind you will have. So take heart in the fact that it will all be worth it in the end and let me continue to explain the wonderful process of making your dream home a reality.

__Highlights of Chapter 5__

- When looking for an area of town, the length of your commute to work should be considered. You and your spouse may want to drive your future commute one morning and afternoon to determine its length.

- The actual length of your commute is not as important as how you personally feel about your commute's length. The amount of time spent on the road may seem longer if you have to drive congested roadways instead of well-designed interstate highways or lightly-traveled back avenues.

- The quality of schools in your area must be checked to ensure your children's education. Even for homeowners without children, quality schools are important for high resale value.

- If your family relies on two incomes or you are a single parent, local daycare centers should be investigated.

- Find what leisure time activities your area has to offer by checking with the local Chamber of Commerce, local websites, or just driving around for a day.

- Discover how far you will be from local shopping so that a quick run to the store does not turn into an ordeal.

- Since visits to church and friends are routine for most of us, the distance we will routinely travel should be kept in mind.

- Even though you may have no intention of ever selling your home, family, work, or other obligations may arise which force you to sell down the road.

- To ensure that your resale value continues to rise, build your home in a growing area where property values are increasing and future development will help raise property values even higher.

- If you currently live in the city where you plan to build, the growing areas of town where you should build are probably obvious to you.

- If you are unfamiliar with the area where you plan to move, ask a local real estate professional for help finding a desirable section of town. Also inquire with the local Chamber of Commerce and tax assessor's office to find average property values, average incomes, etc. and how such growth indicators have been changing in your area over the past several years.

- During your search for a perfect area you will find that the perfect area does not exist. Make a list of things that are most important to you and your family in your new home and keep it in mind as you look for your new area.

- Once you have found your area, drive through local communities and make sure there are empty homesites available for your new home. If not, you will have to find an alternative area.

- Also discover the price range of homes in your area and make sure you can afford them. If you cannot, you will be forced to find another area to build.

Notes

Chapter 6:
Choosing Your Builder

"If a builder has built a house for a
man, and his work is not strong, and
if the house he has built falls in and
kills the householder, that builder
shall be slain."

-Code of Hammurabi
18th Century BC

During a particularly busy period in the mid-1980s, we were building in four local communities. Each community had one or two completed homes built on speculation or "specs" for people who wanted a new home and needed immediate occupancy. In addition to meeting the needs of these clients, our spec homes gave us another couple of models to show along with the furnished model homes located at the front of each subdivision.

Each of our furnished models were manned throughout the day by a site agent or salesperson. A builder's site agent is a very important part of his organization because she is the first person with whom you as a buyer will most likely come into contact. As you go through the builder's model, she will be the one to answer all your questions about the home and the qualifications of your potential builder. Therefore, it is essential that the site agent be informative, give a good impression, and above all be <u>honest</u> when she answers your questions about the home and the builder.

During this time we had placed an ad in the classified section of the local newspaper for new site agents. In response I received a phone call from a young salesman who was working for another local builder. This other builder was building roughly four times as many homes as we were and had a sales staff headed by a man who was famous, or rather infamous, for being somewhat loose with the truth. The young man who called rapidly flooded me with volumes of information on how wonderful his career as a site agent had been with this competing builder. After going through his well-rehearsed sales pitch he closed by saying, "I usually write about ten to twelve housing deals a month."

Considering that most sales agents I have ever dealt with were only able to procure an average of two or three sales a month, even during the best of economic times, I was duly impressed with this young man's claim. Being able to routinely do three or four times better than your fellow workers is amazing in any business, but it is absolutely astounding in the world of sales. However, my amazement began to wane as he followed up his statement with the disclaimer, "Of course, not all of my deals go through. I usually lose about three or four deals a month, but after all, that's as many as six to ten deals a month that do go through."

Shaking my head, I ended our conversation politely by saying I would get in contact with him if a position became available. As

I hung up the phone, the one thought that ran through my head was, "**I WOULDN'T HIRE YOU IF MY LIFE DEPENDED ON IT**." Because first of all, my company does not build deals. And we do not build units. We build homes. Have you ever been invited to your boss' deal for dinner? Have you ever taken your family to grandma's unit for Thanksgiving? Of course not, and you never will. Because although the distinction between the terms "home" and "deal" and "unit" may be just a play on words, I still would feel uncomfortable working with someone who had no more feeling about his work than the term "deal" implies. Anyone who would assist his clients - his future neighbors and friends - to move into a "deal" instead of a home should not be in the business of helping people design and build their dream home. He should instead be trying to sell ice to Eskimos, or some such thing more worthy of his polished selling techniques.

The second thing that would prevent me from ever hiring this young man as a site agent was his statement that three or four of his clients backed out of their contracts each month. What this means is that roughly **1/3** of the people to whom he sold a home had second thoughts and cancelled their agreements. Such a large percentage of people backing out could only mean one thing – he told his clients whatever they wanted to hear in order to sell them a new house. And if what he told his buyers happened to be the truth, that would make his job easier. But if he had to bend the truth a bit in order to get their signatures on the dotted line, that was just an unfortunate part of doing business.

Undoubtedly, being able to write a large number of contracts in a short period of time is a very admirable trait in a salesperson; however, there are other factors that must be considered. You see, the people who did not back out of their contracts with this very pleasant young man spent the entire building process fighting with their builder tooth-and-nail. Fights repeatedly erupted over things that were supposed to be included in the sales price but in fact were not.

No legitimate builder wants to work in that kind of environment. And no well-informed buyer should expect to be told exactly what they want to hear, regardless of the truth. Instead, you need honest answers to your questions and facts about your new home given to you by someone you can trust. If your builder's site agent or representative cannot do this, then this is not the builder in whom you should place your trust.

Over the years, I have been amazed time and time again at the way most prospective buyers choose their builders. First, they are impressed by the friendly site agent who seems so genuinely concerned and caring about what they want in their new home. Second, they are dazzled by the professionally decorated model homes with their impeccably manicured lawns and beautifully coordinated furnishings. And finally, they are impressed because the home has several architectural features that catch their eye such as:

- vaulted ceilings,
- upgraded carpets, and
- Palladian windows, just to name a few.

Too often, people choose a builder for every reason <u>except</u> the one reason that really counts: **THE QUALITY OF YOUR BUILDER'S WORK**.

The thing you have to remember as you look for a builder is buying a new home is not the same as buying other consumer goods, such as a new car. When you purchase your new car, the company that sold the car is not the manufacturer but rather is an authorized dealer for the manufacturer. This means that even if the company from which you bought your car goes out of business, the manufacturer will still uphold the warranty his dealer gave you.

This is not true with your new home. Your builder is **not only** the **dealer** selling you the home, he is **also** the **manufacturer** of your new home. This means that if your **builder** goes **out of business** after he has built your home, you will be left with **no one to fall back on** if your home has any problems down the road.

Whenever I am asked how to avoid substantial construction problems and produce a quality, well-built home, my answer is always the same – **THERE IS NO SUBSTITUTE FOR PEOPLE ON THE JOB WHO KNOW WHAT THEY ARE DOING**. Almost every construction problem that occurs happens because there are not enough knowledgeable, experienced people on the job with years of construction know-how supervising the work. And no amount of regulation passed by government, no matter how well-meaning, will replace someone who knows what they are doing overseeing the construction of your new home.

Unfortunately, most people think that the larger the home builder, the better the quality. Well as we said in Chapter 1 - Myths & Monsters, the truth is that the number of homes a builder constructs has **absolutely nothing** to do with the quality of your new home. Quality depends on the experience and knowledge of the superintendent who is supervising the work on your home. If your superintendent has little construction experience, then you will have a poor quality job; however, if your superintendent has a lot of construction experience, then you will have a good quality home.

At this point, many of you may be asking yourselves, "What exactly qualifies as construction experience?" A construction superintendent should have a track record as an overall supervisor on a construction site. Someone who has done only one facet of the business may be good at that one particular part, but they may not have an overall idea of how a home is supposed to be built. For example, a brick layer who has been laying brick for twenty years may know a lot about how brick should be laid, but he may not know how wood framing is supposed to go together or how mechanical systems (plumbing, electrical, etc) are designed and constructed. Therefore, you need a superintendent who knows how every piece of your home should go together so that your entire home will be built the way it should be.

One of my friends from college who majored in engineering went to work for a paper mill where they had several huge multi-million dollar industrial machines making paper to be shipped all over the world. One day, one of these machines broke down. Four or five mechanics in a row came to the plant and none of them could fix the machine. Finally, the sixth mechanic came in and examined it for a few minutes before taking out his hammer and banging the machine in one spot. Instantly, the machine sprang back to life and the plant resumed full production. The next week my friend forgot all about the mechanic until a bill came for $1,000.

Alarmed, he called the mechanic and asked, "Why did you charge me so much for 10 minutes of work?"

"Would you like me to break the bill down for you?"

"Yes," my friend responded.

Half an hour later, the bill came over the fax machine and read:

Hitting the machine with hammer	$50.00
Knowing where to hit the machine w/hammer	$950.00
Total Due	$1,000.00

Like the mechanic above, only a superintendent who has experience supervising the entire construction process from start to finish will ensure that each part of your home is built the way it should be. As we have said many times throughout this book, and will say again, the quality of your new home is directly proportional to the skill, training, and construction experience of your builder and the superintendent who is directly overseeing the construction of your new home.

At this point, a lot of you are probably saying, "Finding out about your superintendent may be fine once you have a builder in mind, but I can't talk to every superintendent with every builder in town. How do you narrow down a list of builders so I don't have to talk to several hundred jobsite superintendents?"

The first thing you need to do is get a list of builders who are building in your area. The best place to look for this would be the phone book or perhaps a website for local builders in your area. Second, like with all other businesses you want to investigate, call the local Better Business Bureau (BBB) and ask them about each builder to see if any complaints have been lodged against them, how long they have been in business, and if they have had any recent problems that could affect the quality of your new home.

Now at this point I need to add a word of warning about complaints lodged with the BBB. As with other industries throughout the country, the construction industry occasionally falls victim to con artists who use whatever means necessary to steal as much money as they can from legitimate businesses. And the favorite method of these con artists who target the home building industry is to buy a new home from a reputable builder and tell him before closing on the home, "I am going to report you to the BBB (or some other governmental agency) if you do not give me these extra amenities (Jacuzzi tub, bigger driveway, you name it) for free."

I remember one of my fellow builders several years ago who had one of these con artists buy his home and put him through continual agony for over a year. The builder had a great reputation, had been in business for many years, and was one of the best builders I had ever come into contact with during my many years in the building industry. There was nothing at all wrong with the home, and the builder had made every effort to placate his client through numerous trips back to the home to handle the minor callback items that occur with every new home. This particular buyer decided that this was not good enough, and he would milk the builder for all he was worth. For the next several months, the homeowner made bogus complaints to city inspectors, the BBB, and every other governmental agency who would listen. He even had local television news crews filming his home and all of the things that were "wrong" with it. Ironically, many of these "wrong" items that were filmed by the news crews were things that the buyer had done to the home while he was living there and were not caused by the builder at all. Finally, the builder became so desperate to get this con man out of his hair that he offered to buy the house back for more than the original price. The con artist did agree to sell the house back, but he wanted to make enough money from the sale to retire. Inevitably, the case wound up in court where the builder discovered that the homeowner had played a similar con game with a builder in another state. And unfortunately, that con game had caused the out-of-state builder to declare bankruptcy.

So when you are investigating a prospective builder, you cannot rely solely on two or three bad reports filed with the Better Business Bureau. Over a career spanning ten, twenty, or thirty years or more, two or three negative reports may not denote a bad builder, but rather a builder who had the bad luck of running into a con artist that tried, hopefully unsuccessfully, to extort something from him. **In order to find the right builder, you must be willing to dig a little deeper.**

After checking with the BBB, your next step should be to check with the local building department in your new area and ask if they have a Construction Industry Licensing Board (CILB) or a comparable governmental agency regulating their builders. Here in the state of Florida, the CILB is a state agency which licenses and regulates contractors and has the power to suspend or revoke the license of a contractor who proves himself unqualified to build. Consumers who have been the victim of incompetent builders can complain to the board and have them investigate the builder to see if action against his license is necessary. So if your state has a CILB or a comparable governmental agency, you should check with them and see if the builder has had any complaints lodged against his license. If he has, ask what the **findings** of the board were when they investigated the complaint. Like the BBB, the board is also a favorite of con artists, and a builder can be reported to them as well. Therefore, you need to know the **findings** of the investigation into any complaints to see if the complaints were significant. By this time you should be getting an impression of the kind of work your builder has done in the past; however, you still need to take further steps to investigate your builder thoroughly.

While you are prequalifying for your mortgage financing, I would take this opportunity to run your list of prospective builders by your mortgage officer and get his opinion on them. I would ask him specific questions about my prospective builders such as:

- Do you think this builder is qualified?
- Do you think they are competent?
- And most important, would you make them a construction loan[i]?

The reply to these questions in addition to the body language used by the loan officer when he answers these questions can tell you volumes about how he feels when it comes to the builders on your list. If he knows and trusts the builders on your list, he will have no problem recommending them to you. The loan officer will tell you plainly and honestly that you will be very happy with the builders you have chosen. On the other hand, if you mention a builder whom he knows to be incompetent and will give you nothing but trouble, then he will be very hesitant and beat around the bush when he tells you about the quality of the builder's work.

Many of you may wonder just how a mortgage loan officer would know about the quality of a builder's work. Well, when the mortgage company grants a mortgage to a homebuyer who has an incompetent builder, the mortgage loan officers will hear about how shoddy or well the home is built months or even years later. So after listening to several homeowners complain about a certain builder, they eventually know which builders are competent and which builders should be avoided. And if you ask them direct questions about the reputation and the quality of your builders' product, you will be able to infer their opinions with just a little bit of observation.

After leaving your mortgage company, your next step in choosing your builder should be to go into a subdivision where the builder has built recently and talk to the homeowners who have used your prospective builder to build their dreams. This is **by far** your **most reliable source of information** because a homeowner has no stake whatsoever in whether or not a builder gets your business. The only benefit a former client receives from having a builder construct your home is seeing another home spring up in the local area. So if he does not like the quality of his new home or feels that the builder did a lousy job building his new home, you had better believe he will be more than happy to tell you about it.

About ten years ago we were building in a subdivision with four other builders, one of whom had a reputation for poor workmanship. One Saturday I was checking with our salespeople at our model home and a young couple walked in saying they had signed a contract with the disreputable builder, but changed their minds and wanted to buy from us. Naturally I was glad they had chosen us over our poor quality competitor and asked them what had persuaded them to buy from us. They replied that they had been driving around the subdivision and saw a big 4'x8' sign on someone's front lawn that said in 5" high capital letters:

"LET ME TELL YOU ABOUT MY BUILDER!"

So they stopped and asked the homeowner who was sitting on his front lawn what he thought about his builder. During the next thirty minutes, the homeowner frightened them with a long list of things that were wrong with his home. And during his dissertation, he also shocked them by interspersing comments that cast doubt on the marital status of the builder's mother. After their ordeal, the young couple returned to the competing builder, declared that they were no longer interested in buying a home from him under any circumstances, and asked us to build their home instead.

During your interviews with homeowners, please remember that when you find a homeowner who is quick to tell you how bad a job the builder did constructing his home, there is always the danger that you are talking to a con artist like we discussed earlier. Such a homeowner may not be angry with his builder because he received a poorly constructed home; rather, he may be upset because he was not able to cajole or threaten his builder into giving him something he wanted. Therefore, it is very important that you interview **at least** three or four homeowners to make sure you are getting an accurate picture of the builder in question.

Just as a homeowner will be quick to tell you if a builder sells a bad product, they will be just as quick to tell you if a builder constructs a good one. Word of mouth advertising is the cheapest and most effective advertising that exists today, and we have benefited greatly from it over the years. Many times we have had people come in and say, "You built a house for my friend, family member, etc. and they have nothing but good things to say about you. They just love their home, and they told me to be sure to come see you when I wanted my house built, so here I am."

In fact, we have gone so far as to gather some of these homeowner comments about our company and make them part of our promotional items so that people can see what our former clients think of us. And of course, our former homebuyers stop by our model center from time to time and socialize with our salespeople; so whenever a potential client happens to visit during these times, our former homebuyers become our most effective promotional tools.

But no matter how big or small the size or how good the reputation of a business, there is always the danger that a company may hit upon hard times. Even the best and the brightest have gone bankrupt; so you need to make sure that your builder is not in immediate danger of doing the same. The best way to do this is to get a credit report on your builder and make sure he has a good credit rating as well as a fine reputation. Some people may feel awkward about checking the credit of someone because they feel it is somehow invading their privacy. But when you consider that the mortgage company is going to check your credit to make sure you can pay your bills, you should make sure your builder can pay his.

After several weekends of visiting builders' models and researching your builders to exhaustion, you will have by this time whittled your list of prospective builders down to two or three who all have the following:

a)   technical competence and years of training and experience
b)   the ability to do customized design work
c)   a good reputation that has been established over time in the community

Now you are ready to go one step further in your research and actually meet the superintendents for each builder who will be overseeing the construction of your new home. Ask your sales representatives to set up an appointment with the superintendent who will be responsible for your home and ask him about his background, his training, and how much experience he has as a jobsite superintendent. Remember, the quality of the superintendent overseeing your home's construction is directly proportional to the quality of your home. If your superintendent was pumping gas at the corner station a month ago and decided to get into construction because he believes it is the path to fame and fortune, you should run, not walk, as fast as you can to a competing builder.

Also, ask your superintendent how many jobs he tries to manage simultaneously. The number of homes a good superintendent can keep track of will be limited by factors such as: how far apart the homes are located from one another, the size of each job, and the complexity of each job. For example, it is a lot easier to manage fifteen starter homes that are all side-by-side on the same street than manage two multi-million dollars custom mansions located on opposite ends of town. If your superintendent has years of construction experience and is not trying to supervise more than he can handle, then you should feel comfortable knowing a skilled professional is guiding your home's construction process.

By the way, I know most of you are wondering just how many homes being managed by a superintendent constitutes more than he can handle. And the easy answer is – I cannot tell you. Trying to come up with a number for every superintendent with every company is like trying to come up with a pick-up line that will work with every woman or man you will ever meet. It simply depends on each individual superintendent in each individual building company. For some, trying to manage one starter home is one too many, while for others who have good people in their company to rely on, ten multi-million dollar mansions spread all over town may be managed with no problem at all. When you meet your superintendent and ask him questions about himself, you should be able to infer what kind of a person he is and therefore what kind of a manager he is. With this knowledge and the answers he gives you during your interview, you can make an educated guess as to what kind of manager he is and whether he will do a good job for you.

Your final step in choosing your builder (I'll bet you never thought we would ever get here) will be to gather your list of remaining builders and revisit their model homes. During these return visits, you need to look <u>very seriously</u> at the homes. When I say look at the homes, I do not mean look at the drapes, the furnishings, or whether the décor is color coordinated. After all, you will decorate your own home and what you place in the home may be drastically different than what the builder's interior decorator installed in his model.

Instead, I want you to look at the **quality** of the home itself. Look at how the trim work was installed around the doors and along the walls. Examine the interior and exterior painting for blemishes or sloppy work. Inspect the roofing, the cabinets, the flooring, and literally nit-pick the model to pieces and see if you would accept the same quality workmanship you find here in your next home. The model home is the best your builder can do – he would not put his second-rate work on display for the public to examine. Therefore, if your builder's best is not good enough for you, then you will <u>never</u> be happy with any home this builder would construct for you.

Once you have done all the steps in this chapter, the choice of a builder should become self-evident. Once you have checked your builder thoroughly and found his reputation, credit rating, and the quality of his work to be excellent, you should feel comfortable in the knowledge that your builder is going to deliver an excellent home. Some people will tell you that researching your builder so in-depth is a waste of time, but a few years down the line you will find that some of these are the same people who are telling you horror stories about their new homes. So doing a little homework on your builder now could save you a lot of headaches further down the road.

Now, I realize that probably less than one in a thousand buyers will do anything approaching the amount of research we are calling for in this section. Most people will go in a builder's decorated model and make a snap decision with regard to their home purchase. And unfortunately, a lot of them are going to make the wrong decision. **The sad reality is that even if you do all of the things that we discuss in this chapter, there is still no guarantee that you will make the right decision.** However, in any situation where you invest a large amount of the savings that you and your family have worked so hard to build up, **<u>you will always be better off making your decision based on knowledge</u>** instead of ignorance. I know most will make a snap decision when they buy their dream home because the court dockets are full of cases where people have tried to make their builder pay for their rashness. Hopefully after having read this section, you will not become one of them.

__Highlights for Chapter 6__

- As a well-informed buyer, you should not expect your builder to tell you exactly what you want to hear. Instead, you should expect honest answers to your questions, given by a builder you can trust.

- Much too often, people choose a builder because of the friendly site agent, the home's decorations, or an ornamental architectural feature that catches their eye. They choose a builder for every reason except the one reason that really counts: **THE QUALITY OF YOUR BUILDER'S WORK.**

- Unlike other major purchases, your builder is not only the dealer but also the manufacturer of your new home. Therefore if your **builder** goes **out of business**, you will be left with **no one to fall back on** if your home has problems.

- The quality of your new home depends on the construction experience and knowledge of the superintendent directly overseeing your home's construction. Construction experience consists of having experience as an overall supervisor on a construction site.

- After obtaining a list of local builders from the phone book or a website, call the local Better Business Bureau (BBB) and see if any complaints have been lodged against them, how long they have been in business, and if they have had any recent problems.

- Check with your local building department and ask if they have a Construction Industry Licensing Board (CILB) or comparable governmental agency regulating their builders. See if your builder has had any complaints logged against his license, and the **findings** of the investigating Board.

- Ask your mortgage officer if your builder is qualified, competent, and would he make him a construction loan.

- Go into a subdivision where your builder has recently built and talk to homeowners who have recently used your builder. This is **by far** your **most reliable source of information on your builder**.

- Get a credit report on your builder to ensure that he has a good credit rating as well as a good reputation.

- Meet with the superintendent who will be responsible for your home and ask him about his background, training, construction experience, and how many homes he tries to manage simultaneously.

- Revisit your builders' models and look at the **quality** of the home itself. Inspect the home and see if you would accept the same quality workmanship in your home.

[i] construction loan – a loan similar to a mortgage that a bank grants to a builder when he is building a home. This loan is repaid to the bank by the builder when the home is sold.

Notes

Chapter 7:
Choosing Your Homesite In A Subdivision

When given a choice on where to build their dream home, most people prefer a homesite in a subdivision instead of a plot of land in the middle of the country. The reason for this is simple – a well-designed, modern subdivision is by far the easiest, most convenient, and almost always the most economical place to build.

Subdivision homesites come complete with good roads and all the utility services, such as water, sewer, and cable TV, necessary for modern living. Therefore, when you build in a subdivision, your builder does not need to tame the savage wilderness to provide a shelter for you and your family. Instead, he only needs you to pick the homesite that best fits your needs and stand back while he builds your dream.

Sounds simple doesn't it? Well, like so many get rich quick schemes you get bombarded with everyday, this process of picking a lot in a subdivision is more complicated than it sounds. Before you can even begin to look at your new potential neighborhood, you first have to become aware of the surrounding area. This way, you can avoid potentially large problems before they grow big enough to inconvenience you. For example, although you may love your new home, eventually there will come a time when you want to leave. You may feel this desire because you feel compelled to get away for a while, you ran out of groceries, you need new clothes, or you have to go to work because your rich uncle did not leave you a dime and the mortgage is due. When you decide to depart, you will want to get in and out of your neighborhood as quickly and trouble-free as you can. So while you are searching for a new neighborhood, keep this in mind and ask yourself the following questions:

❑ **How many entrances does my new subdivision have?**

Remember that the more entrances you have, the easier it will be to come in and leave the subdivision. However, with this added convenience comes the increased danger of people who are not residents driving through your subdivision as a shortcut to get to another place. This increased traffic could be a source of concern, especially if you have small children.

❑ **How many people will be using the entrances to your subdivision?**

The more people you have in your subdivision, the more cars will be using the entrances to your subdivision and the longer it will take for you to leave and enter your neighborhood.

❑ **How busy are the highways these entrances face?**

With more cars using the roads around your community, you will have a harder time getting to and leaving your home. Therefore, finding the amount of traffic around your subdivision during the peak times you will be leaving and entering (going to and from work, for example) would be a good idea.

Another item to look for in the surrounding area is the location of shopping centers. Having stores and marketplaces nearby will give you the benefit of convenient shopping on the weekends and after work. However, other people who want the same convenience may cause a large amount of traffic to pass by your neighborhood during peak shopping hours. This traffic may not seem like a big deal if you are on your way to your favorite weekend fishing spot or going to see your in-laws, but if you overslept and wound up running late to work one morning, these traffic delays could become serious very quickly. And if you decided that convenient shopping meant having the corner store located literally on the corner adjacent to your neighborhood, you may find yourself the victim of some sleepless nights in the years ahead. You see, in some areas, store delivery trucks can start their deliveries as early as four in the morning and act as the perfect alarm clock – set three or four hours before you need it.

Like a shopping center, having a big wooded area next to your subdivision can also be a benefit as well as a detriment. Big trees and full shrubs will give your neighborhood a beautiful rustic charm that you will appreciate as you wind down from your busy work day. The woods will also provide a nice play area for kids away from the dangers of traffic and strangers. However, you may have to face the prospect that one day this wooded area will disappear and be replaced by a new building. Almost every time I have built on a leftover or "infill" lot in an older subdivision, I have had people call me and swear that when they bought their home, their sales representative promised that this beautiful, wooded plot of land would never be touched. I have seen homeowners yell, curse, threaten to sue, and even threaten myself and fellow builders with bodily harm when they saw bulldozers sculpting the pristine landscape. And every time a situation like this has come up, I have told these misguided homeowners the same thing I am about to tell you.

The only way to make sure that a plot of land will remain undeveloped forever is to buy the land yourself and not allow anyone to build on it. Because no matter how slowly or quickly an area grows, it will never, I repeat **NEVER,** remain forever like it is today. Inevitably, given enough time and an increasing population, someone will buy a rural parcel and develop it to accommodate the growing number of people. As more and more people develop land in this way, the rural areas of town will evolve into a modern cityscape. Therefore, the only way to keep someone from developing land in a growing area is to beat them to the punch and buy it yourself. That way, if anyone builds anything on your pristine land, you have no one to blame but yourself.

This evolution of land from undeveloped, unspoiled wooded acreage to modern developments is nothing new. In fact, if you look at any area located "close in" to a city or town and research the area, you will find that these developed areas were in fact at one time considered rural. A perfect example of this is the old Ortega section of town here in Jacksonville where I grew up. When my father moved our family there, it was a fairly new housing development on the very edge of town. I still remember my friends' fathers and grandfathers telling me about how they used to hunt and fish all over the area where my house stood. Fifty years later, that same house that was considered to be on the very edge of town is now considered to be in the middle of town. Thousands of people, including those we build new homes for today, have to drive through the area of town where I grew up on their commute to work. And I know that one day some of the areas where I am now building that are considered on the "edge of town", will one day be considered "in the heart" of the city.

So no matter what anyone says, you should logically assume

that vacant areas of land will one day be replaced by some kind of development. One day, that beautiful group of oak trees standing majestically on that undeveloped parcel at the end of the street may evolve into a home for your future friends and neighbors. Before you choose an area to build, realize that the landscape around your home may change as the years go by. And as the scenery changes, make sure your reason for living there does not change with it. Because if it will, you need to examine your choice of neighborhoods more carefully.

In addition to wooded areas, other subdivisions in the area can also become a source of concern if you do not look at them before you buy your new home. One of the more obvious concerns is the fact that with an increase in the number of people living around you, there will be more people on the road trying to get from one place to another. This increased traffic will not benefit you in the least or make your property values go up. However, if the homes in your surrounding subdivisions are less valuable than yours are, these people can actually harm you by dragging the value of your home down. Having people both in your way and sinking your property value is no way to live, so look at the subdivisions surrounding your home before you decide to build there.

Speaking of other subdivisions, you may discover that once you have found a place you want to build, your builder may not be able to build there. This is not because he is incapable of building your home or does not want to build there, but rather he cannot obtain a lot in that area. You see, when a subdivision is developed, a builder usually goes to the developer and reserves a certain number of lots in the development for his new homes. Thus once the neighborhood is under construction, all of the homesites have usually been reserved by the builders who already build there. This is why you may find that although you want to be in a certain neighborhood, your builder may not be able to purchase a homesite from the developer if he is not currently building in that area.

When you have a situation like this arise, you usually have three options:

1) buy a home from a builder currently building in your chosen area

2) buy a home from your builder in another subdivision, or

3) build your home on an "infill" lot

With your first two options, you already know exactly what to do because you have been carefully reading and memorizing every word of this book as we went along. (However, if you are like me and sometimes have a few details slip by you when you read something, please review Chapter 6: Choosing Your Builder and the beginning of this chapter.) However, with your third option of "infill" lots, I need to give you a bit of an explanation.

An "infill" lot is a vacant homesite or a vacant piece of land located in an area that has already been developed for some time. Such lots may be in the middle of downtown, an old section of town, or in an existing subdivision – anywhere that substantial development has already occurred and the land surrounding the lot has been developed. When you find a new subdivision but discover that your builder cannot build there, you may drive around the area and find an equally desirable subdivision that has a few undeveloped lots remaining. Some people may think that because these lots were not taken, there must be a problem with them. Although sometimes this may be true, usually it is not. Instead, the lots are not developed for other reasons such as the builder who reserved it went broke and the developer never resold it, or the developer may have gone broke and the bank repossessed the lot. And in some cases, a person like yourself who wanted to build their dream home there bought this particular homesite and has simply not built on it yet.

A couple of years ago, a friend of mine ran into a situation like this while locating a homesite for herself. Driving around the area, she passed through an existing subdivision and found a vacant, wooded lot surrounded by ten-year-old homes. After searching through the tax records at the Property Appraiser's office, she discovered that the lot was owned by a couple in Georgia who were planning to build their retirement home there but had since changed their minds. Having been away for ten years, they did not realize how much their property was worth; therefore, when they were offered $10,000 for their land, they took it in a heartbeat. My friend was able to get this $25,000 lot for $10,000 and apply the remaining $15,000 toward some great extra features in her new home.

In addition to the occasional lucky opportunity like my friend stumbled upon, there are many other advantages to building on an infill lot. For starters, you know exactly what will be built around you because it is already in place. There will be no chance of waking up one morning and seeing something completely undesirable being built right next to you because everything that could be built around you is already there. Driving through an existing neighborhood, you are able to see if the homes and the lawns are well-kept. This can be extremely important not only for your own comfort and peace of mind, but also for the comfort of a future homebuyer when it comes time to sell your home. And being one of the last homes built in the community, you get to meet your neighbors before you move in. This can also be a decided advantage when you consider that these are the people who you will see everyday and will watch over your home when you are away.

So with these advantages in mind, you may want to add infill lots to your list of possible subdivisions for your next home. Existing subdivision homesites do require more legwork to find because they will not be as aggressively advertised as sites in new communities. In fact, they may not be advertised at all. But with a little bit of work and some patience, you may come to realize that a subdivision does not have to be new to fit your needs.

After carefully selecting the subdivision you want, you are now ready to start selecting the site of your new home. When you visit your builder's model home, your sales representative will give you a diagram of the subdivision and show you which homesites are still available. This diagram is called a **sales plat** and shows the layout of the homesites and the roads in the subdivision and identifies each homesite by a **lot number**. These sales plats usually do not have the size of the lots printed on them or any kind of dimensions. They are only meant to give you a rough idea of your neighborhood's size and layout.

County Road 220
The Glades Road
Redwood Lane
Lake
Lake
Nature
Preserve
1
2
3
4
5
6
7
8
9
10
11
12
13
14
15
16
17
18
19
20
21
22
23
24
25
26
27
28
29
30
31
32
33
34
35
36
37
38
39
40
41
42
43
44
45
46
47
48
49
50
51
52

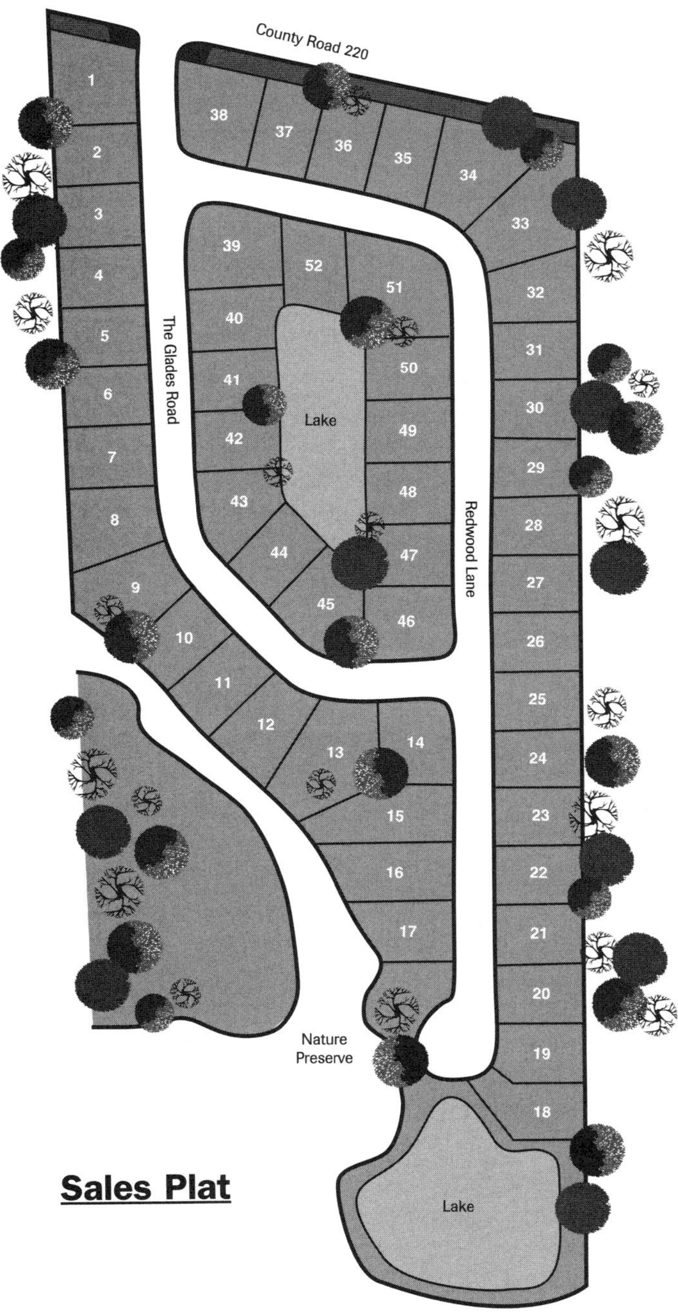

Sales Plat

With your sales plat in hand, you are ready to whittle down your choice of lots to a few select homesites. As with everything in your life, the first concern should be for the safety and well-being of your children. In order to make sure that the traffic on your street both now and in the future does not look like an interstate highway, you need to ask the following questions:

➢ Is your street a main thoroughfare? If it is, will this allow people to use your street as a shortcut to get to other places instead of stopping in your subdivision?

➢ Will there be future development tying your part of the subdivision to another phase of development, thereby increasing the traffic passing by your home?

➢ Will a connection be made at a current dead end road tying that road into another subdivision, thereby increasing the traffic count substantially over the next several months or years?

If you answered "yes" to any of the above questions, chances are that the nice, quiet street your home faces will evolve into a busy thoroughfare in a short amount of time. Not only will the increase in the number of vehicles add unwanted noise, but it could also be potentially dangerous to small children who are too young to know the danger of a speeding car. Doing a little bit of homework, or "roadwork," beforehand could save a lot of headaches and heartaches in the years to come.

While you are checking the roads for potential hazards, take some time to talk to your future neighbors. Since they already live where you want to start your home, they can provide some valuable insights into the area.

And if your children are anything like mine were, they will undoubtedly be playing in the neighborhood after the sun goes down. To make sure they will not be playing in the dark, visit the neighborhood at night and see if the street lighting is adequate. Too little light may contribute to the decline of the neighborhood down the road because, as most police officers will tell you, a crime-ridden area is rarely well-lighted. On the other hand, too much lighting can also be a problem. You do not want a street light shining in your bedroom window at 3 A.M. while you are trying to get some sleep for that business meeting at 7 A.M. the next morning. Since everyone has a different opinion of how much is too much, you should drive and/or walk through your new neighborhood to see if the lighting is right for you.

Another source of annoying lighting that few people think about is car headlights. If the lot you choose is positioned so that the street dead ends directly into your home, you may find that every car coming down the street shines its lights directly in your front windows. This can be quite irritating, especially if the master bedroom has a large front window that allows any passing headlight to splash across your face whenever you get into bed. Sometimes you can fix this problem during the design phase by placing rooms that are seldom used on the front of your home and positioning the bedrooms on the back. This way, car headlights will not even be noticed. However, you may find it easier to solve this problem by choosing another lot where you do not have to deal with passing headlights at all.

Once you have made several visits to the available homesites and narrowed down your list of choices, it is time to take a closer look at each selection. When a developer builds a new subdivision, he hires a Professional Engineer (PE) to divide the land into individual lots or homesites for sale to the public. Once the PE has designed the utility lines and the layout of the streets, he makes a **record plat** and records it with the county government. As you can see from the record plat of our community the Glades at the back of this chapter, a record plat shows the property lines for each lot, their length and orientation (compass direction), precise measurements for the layout of the streets, and the easements which may or may not lie along property lines. Each of these characteristics must be carefully considered as you look at each homesite because they can make the difference between a livable house and a comfortable home.

When you examine the property lines for each homesite, keep in mind how large the home you want to build is going to be. This way, you can know if the home you want to build will actually fit on the lot or if a bigger homesite is required. While you are fitting your home on your lot, realize that local zoning laws in your area require you to maintain a certain **setback** distance from each property line. These setback distances vary from location to location, but all will dictate that you must stay a certain number of feet from the front, rear, and side property lines. For example, in our local subdivisions our models are required to be 20' from the front property line, 15' from the rear, and 7.5' from the side property lines.[1] Now your home can be located farther away from these property lines than the minimum distances, but it cannot be any closer.

While you are picturing your home on each lot, keep in mind how large you want the front and back yard to be. You may find you want a large backyard if you are thinking of getting a swimming pool installed after you move in. However, if you do not particularly like to work in the yard or have a "yellow thumb" when it comes to gardening, then you will definitely want a homesite with minimal yard space.

During your examination of potential homesites, you may find that a few show drainage or utility **easements** marked with dotted lines adjacent to the property lines. Easements give someone the right to come onto your property and do work within the easement area if it becomes necessary. One example would be a drainage easement on your back property line which would allow the local government to come and cut swales or small ditches in this easement if the drainage in your area needed to be reworked. Another example would include a utility easement on your front property line that allowed local utility companies to come and service underground electrical, telephone, and cable TV lines located in this easement if necessary.

In the past, some clients have been concerned after seeing an easement located on their property line. What they failed to realize was that almost 99 times out of 100, these easements are never used. Easements can simply be thought of as an insurance policy, protecting you and your family in case a problem arises with your neighborhood drainage or utility service. The only concern you as a homeowner should have about easements is whether or not there is enough room to fit your home on the lot when the easements are present. You see, nothing can be built in an easement that a crew of men could not haul or move away. So you could not build a home or an in-ground pool or a garage in an easement because these are permanent structures. However, if you wanted to erect a fence in an easement, that would probably be allowed because it could be easily removed and put back when the easement work was completed.

One final thing you need to be aware of when you look at your record plat is the orientation of your new home. The direction your home faces has a lot to do with how livable your home is depending on the home plan you choose to build. For example, if your kitchen faces the rear of the home, you will want the back of your home to face east if you enjoy the sunlight streaming into your kitchen in the morning. However, if you are like me and think that the best thing about the morning is that it is over by the afternoon, then you will not want the sun to slap you in the face while you are eating breakfast. For people like us, our kitchen should face anywhere but east so the sun will not bother us as we awaken from the dead. Something like the orientation of your home may not seem like a large thing before you build, but it can certainly become huge after you move in.

In the real estate industry, there is an age old adage which says that the three most important things to consider when buying property are "Location, Location, and Location." When you are selecting a place to build your new home, there are three things to consider that are even more important: "Drainage, Drainage, and Drainage," in that order.

Many people make the mistake of picking their new homesite on a beautiful, sunny day when birds are singing and the neighborhood children are riding along on their bicycles, enjoying the midday air. People forget that soon after they move in a rainy day is coming, especially in the tropical climate of Florida where we work. And when it comes, the stormwater could decide to get up close and personal with you – sometimes by inviting itself inside your home and making an indoor swimming pool.

I remember one time about fifteen years ago we were building in an area where a competing subdivision was located directly across from us on a busy thoroughfare. The homes in the competing subdivision were built quite a bit lower to the ground than ours were – so low you could tell the difference just by looking across the street from our community. That summer we had a torrential rainy season that left many areas deluged with rain. One Saturday while I was eating breakfast, I turned on the TV and saw something I will never forget as long as I live. Plastered across the screen was my competitor's model home sitting in about two feet of standing water. Stunned, I sat there watching as a homeowner in a canoe rowed past the local reporter and held up some groceries he was taking to his marooned family. I joked with one of our salesman that I should give our competitor a price for building a drawbridge after he figured how long he needed it. But I quickly decided that such an offer of goodwill might be misinterpreted.

In a properly engineered subdivision, the design engineer has developed a neighborhood **drainage plan** to ensure that stormwater <u>never</u> has a chance to get up close and personal with you. To determine how your lot and others around it are supposed to drain, you, as an informed buyer, would not be out of line to ask to see a drainage plan. In more than 30 years of building and over 1,000 customers, there has **<u>never</u>** been **one** person that asked to see a neighborhood drainage plan. Not **one**. If I were having a house built and investing hundreds of thousands of dollars of my money, I would ask to see a drainage plan. In fact, I would not buy a new home unless I could see one. As an informed buyer, you should feel the same way.

When your builder shows you the neighborhood drainage plan, there are several things you need to look for. As you can see from the drainage plan of our subdivision <u>the Glades</u> at the end of this chapter, the neighborhood is divided into several parts by heavy, dashed lines. These lines are called Drainage Breaks and separate the neighborhood into drainage areas. All of the water which falls into one of these drainage areas is routed by the contour of the land to drain in a certain direction.

The direction the water will drain is determined by the design engineer and is shown as "A", "B", or "C" drainage. Lots in an "A" drainage area will be graded so that all of the water falling onto that lot will drain to the front of the lot and into the street.

The elevation of the rear property line is higher than the front line. The ground slopes down from the rear property line to the front. The water follows this slope and flows from the back yard to the front yard, over the curb, and to the storm inlet or drain.

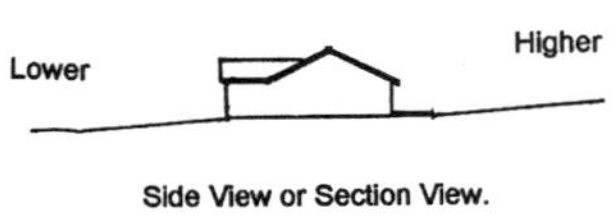

Side View or Section View.

The lot is graded so water from the back yard flows around the house and into the street. A swale or shallow ditch is dug around the house while landscaping to channel rainwater around the house.

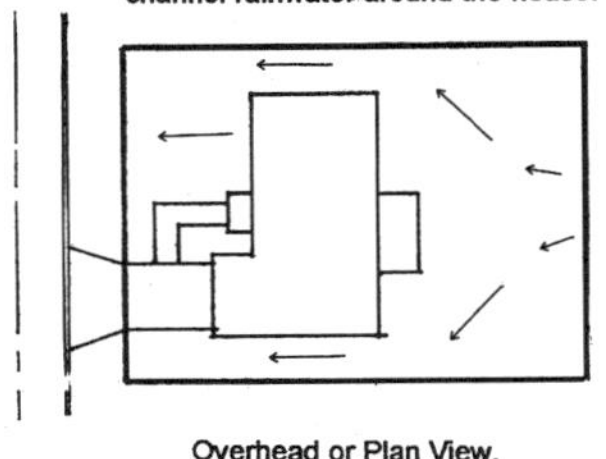

Overhead or Plan View.

<u>"A" Drainage</u>

The center of the property is higher in elevation than the rear property line or the front line. The ground slopes down from the center of the property to the rear and front of the lot. The water follows this slope and flows from the center of the lot where your home is located to the front yard or back yard. From there, the water follows the subdivision drainage plan and finds its way to a storm inlet or drain through the front yard like "A" drainage or through swales or small ditches on the back property line in the back yard like in "C" drainage.

The elevation of the rear property line is lower than the front line. The ground slopes down from the front of the lot to the rear property line. The water follows this slope and flows from the front yard to the back yard, into a swale or small ditch on the rear property line, and to a storm inlet/drain or a stormwater retention pond where it evaporates.

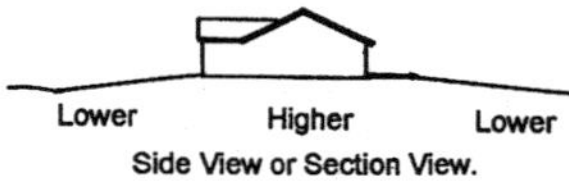

Side View or Section View.

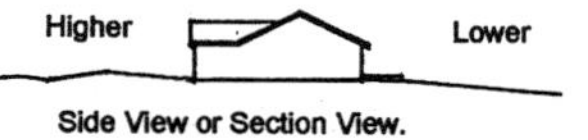

Side View or Section View.

"B" Drainage

The lot is graded so water from the front yard flows around the house and into the swale at the rear property line. A swale is dug around the house while landscaping to channel rainwater around the house.

Overhead or Plan View.

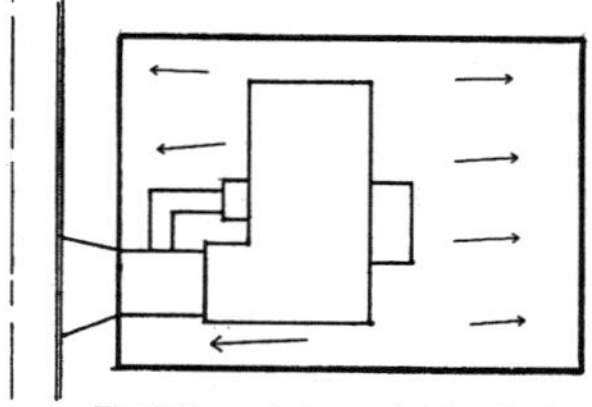

The lot is graded so water from the back yard flows across the back property line and into a swale or small ditch and water from the front yard flows into the street. A swale or shallow ditch is dug at the back property line while landscaping to channel rainwater off the back property line into a drainage inlet or retention pond, depending on the neighborhood drainage plan.

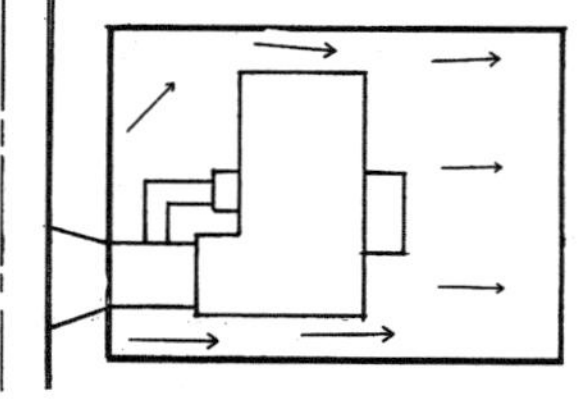

Overhead or Plan View.

"C" Drainage

"B" drainage lots are graded so that water that lands on the front part of the lot facing the street drains to the street while water that hits the back part of the lot drains off the back property line. Finally, "C" drainage lots are graded so that all of the water drains from the front of the house to the back and is carried off the lot at the back property line.

Another important thing to watch for when picking a lot is the number of trees that will remain once your home is built. I remember one time when we built in a subdivision that was formerly an old dairy farm. The rolling pastures that made nice grazing land for cattle also made good draining soil that worked well for building. The only problem was that trees and pasture do not mix; therefore, the land did not have the lush trees most people wanted in their neighborhood. This caused me to lose a number of buyers who wished to live on wooded lots, but was much less of a headache than what one of my competitors had to endure.

In his subdivision, my fellow builder had nice, undeveloped wooded lots that people fell in love with at first sight. What they did not realize was that these lots required a lot of fill dirt to raise them to the finish elevations the engineer called for on his drainage plan. And trees and dirt go together like plaid and stripes. Excessive fill dirt around trees will smother them to death; therefore, trees have to be removed when a lot of excess dirt is required. So when these people who hand-picked their wooded lots came back to see their completed dream home, they found their dream planted in the middle of a barren wasteland instead of tucked away in a wooded glade.

Needless to say, the people who found their trees stripped away were not happy and were a constant source of headaches for my fellow builder. In order to make sure that your wooded lot does not become an empty field before your eyes, you need to do some preparing before you pick your new homesite. Go back to your neighborhood drainage plan and look at the finish floor elevation for your homesite. Compare this elevation to the elevation or contour lines for your lot. If the finish floor elevation is only 1 to 1-1/2 feet above the nearest contour line, then you will probably need a minimal amount of fill dirt and most of the trees on your homesite can be saved. So as you look for a new homesite, please remember that if you want to keep most of your trees, you will need to choose a lot that does not require an exorbitant amount of fill dirt for your new home.

After reading this chapter, those of you who want a homesite in a subdivision should now have more than enough information to make an informed choice. However, for those of you who need more "breathing room" than an engineered neighborhood allows, the next chapter on choosing your homesite outside of a subdivision was written especially for you.

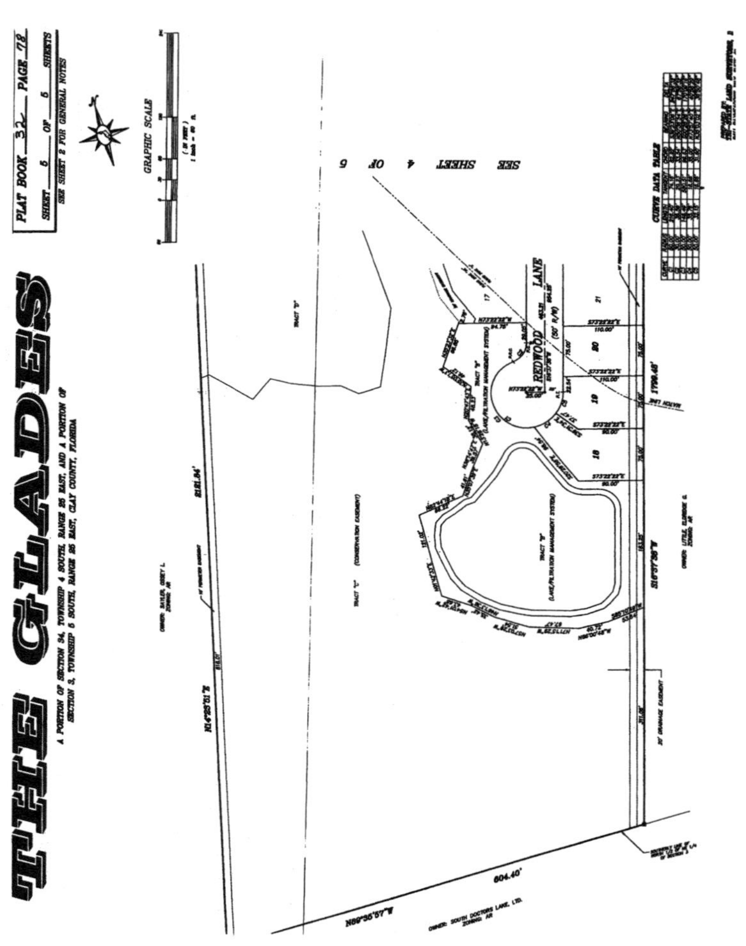

Record Plat

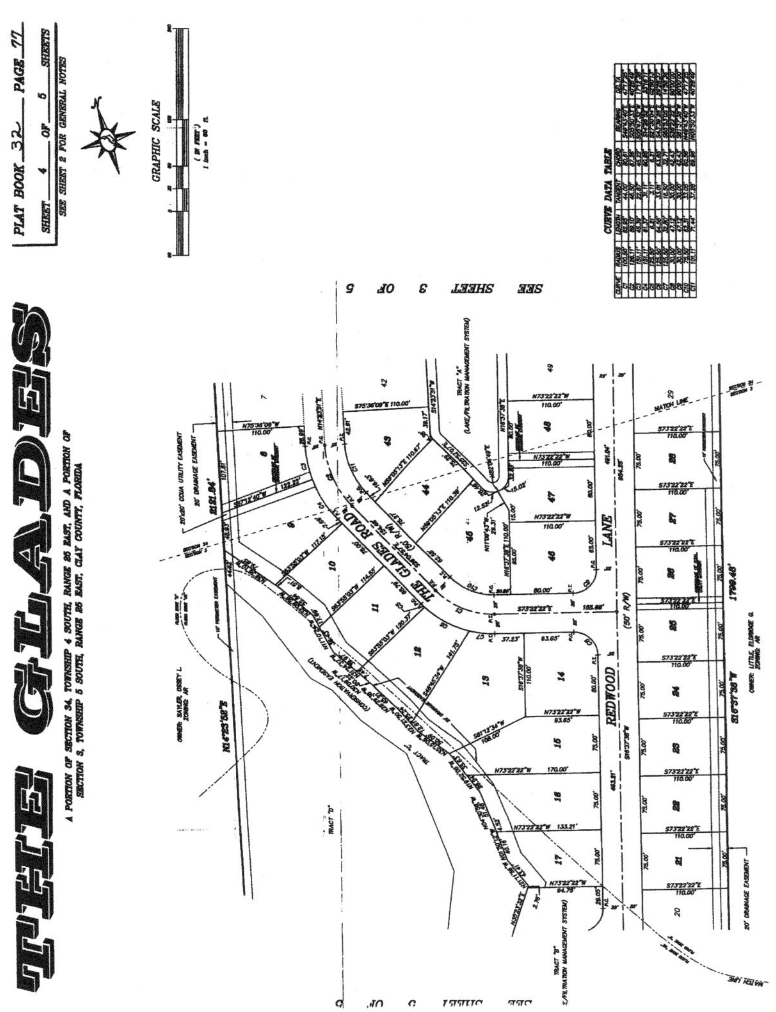

Record Plat

THE GLADES
A PORTION OF SECTION 34, TOWNSHIP 4 SOUTH, RANGE 25 EAST, AND A PORTION OF
SECTION 3, TOWNSHIP 5 SOUTH, RANGE 25 EAST, CLAY COUNTY, FLORIDA
PLAT BOOK 32 PAGE 76
SHEET 0 OF 5 SHEETS
SEE SHEET 2 FOR GENERAL NOTES
GRAPHIC SCALE
(IN FEET)
1 inch = 80 ft.
CURVE DATA TABLE
COUNTY ROAD NO. 220
(FORMERLY STATE ROAD NO. 220 – 100' R/W)
POINT OF BEGINNING
POINT OF REFERENCE
REDWOOD LANE
THE GLADES ROAD
TRACT "A"
(LAKE/FILTRATION MANAGEMENT SYSTEM)
MATCH LINE
SEE SHEET 4 OF 5

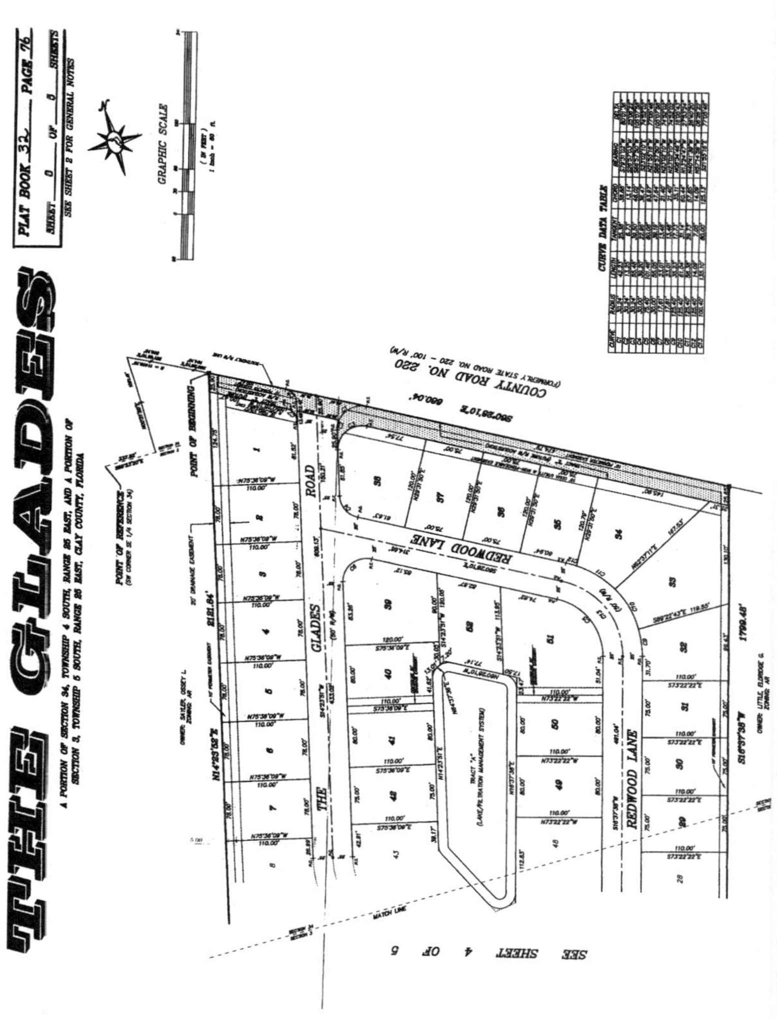

Record Plat

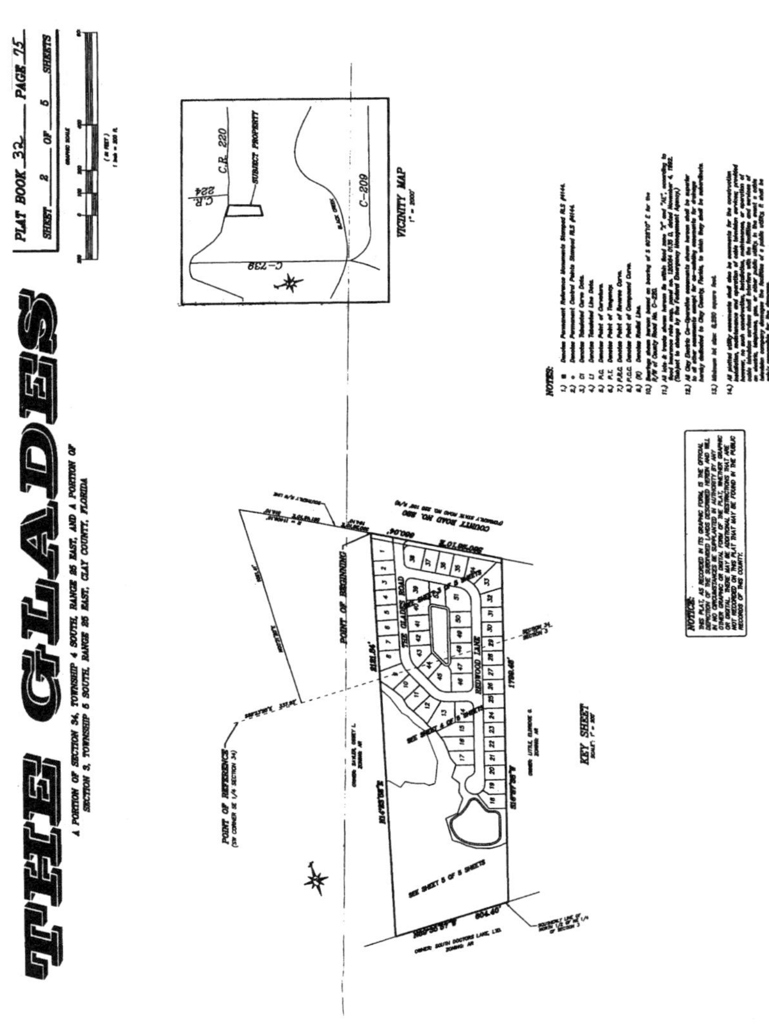

Record Plat

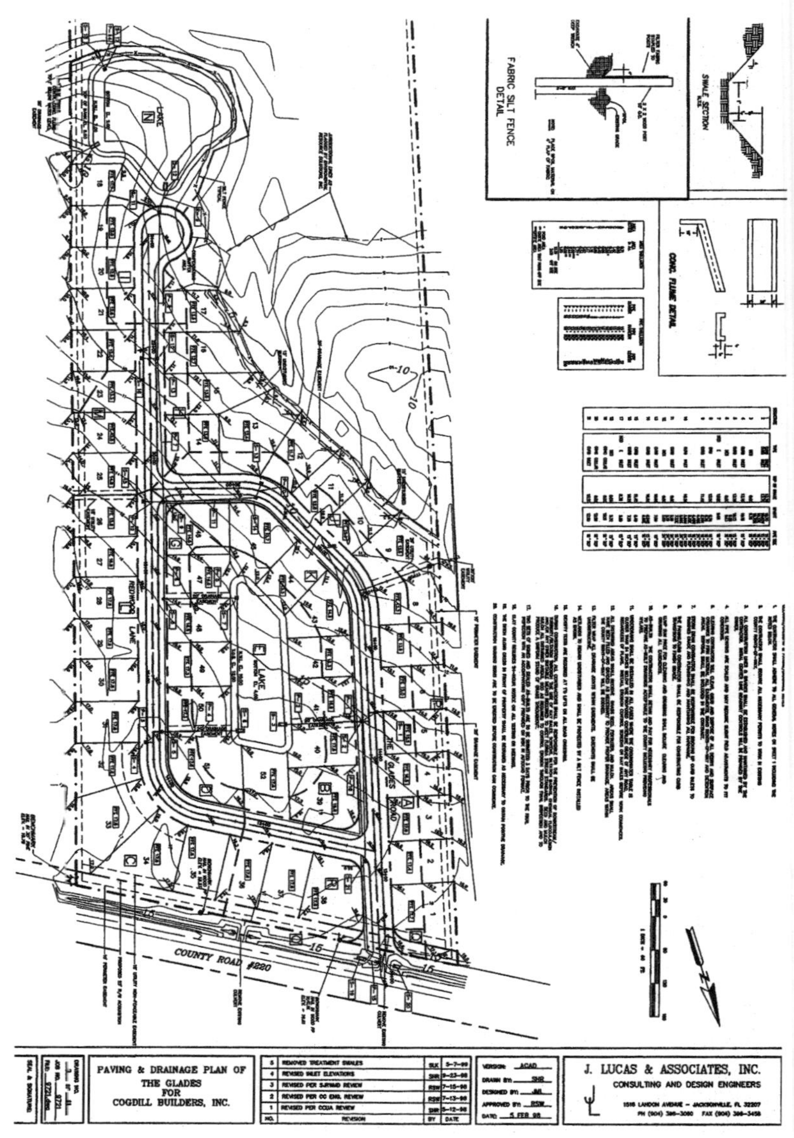

Drainage Plat

Highlights from Chapter 7

- Find the number of entrances your subdivision will contain. The greater the number, the easier coming and going will be, but the more traffic you will have to contend with.

- Discover how many homesites your community will have when it is finished. The more neighbors you have, the more traffic you will have and the longer it will take to enter and leave your subdivision.

- Ask about the traffic on the roadways around your community during peak hours. The busier the roadways, the harder time you will have entering and leaving your community.

- Pinpoint the location of local shopping. Having marketplaces nearby means convenient shopping but may also translate into heavier traffic during peak shopping hours.

- Wooded areas are attractive and provide a nice play area for children. Just remember that any rustic areas around your home will most likely be developed and new buildings put in their place.

- Surrounding subdivisions can lead to an increase in traffic and can also drag down the value of your home if the homes built there are less valuable than yours.

- When you discover an attractive community, your builder may not be able to acquire a homesite there. In this case, you may have to buy your home from another builder currently in that subdivision, buy your home from your builder in another community, or build your home on an "infill" lot.

- "Infill" lots are vacant parcels of land located in an area that has been developed for some time. Although sometimes these lots have problems, in most cases they are usable and can be a bargain.

- "Infill" lots allow you to see what is built around you before you move in, see if the neighborhood is well-kept, and meet your neighbors.

- Check drainage plans to ensure water does not turn your family room into an indoor swimming pool.

- Inquire as to how much fill dirt will be required and how many trees can be saved before you choose your homesite.

- Remember drainage, drainage, drainage in that order.

[i] Setback distances can vary depending on zoning and deed restrictions in your area.

[ii] Record and Drainage Plats of "The Glades" done by Jim Lucas and Associates, Inc. of Jacksonville, Florida. Used with permission.

Notes

Chapter 8:
Choosing Your Homesite
Outside Of A Subdivision

Many times throughout the year I receive phone calls from people who are thinking of building on their own land. These future homeowners have usually found a parcel of land a little further away from town, fallen in love with it, and have negotiated a price a little below market value with a very anxious seller. And now that they have made a great deal on their homesite, they are excited because they are absolutely certain their dream home can be built for considerably less than building inside an engineered subdivision.

After all, when builders price a home in a subdivision, that price usually includes the price of land. So obviously if the land is already paid for, you can subtract the price of the land and build your home for less, right? **WRONG!** Unless the property owner has inherited the land or bought it for much less than its market value, chances are it will **cost more, not less** to build on that site than in a modern subdivision.

One reason for this surprising fact is a builder has his supervisors, material deliveries, and subcontractors set up to build multiple homes in his subdivisions. As Sam Walton proved with the success of his Wal-Mart® stores, goods and services can be bought for less when you buy them in volume. Therefore, since it will cost your builder more money to reset his operation and build on your lot for a "one shot" deal, he will pass his increased costs onto you.

Second, as we mentioned in the last chapter, a modern subdivision is designed by professional engineers to provide all the infrastructure necessary for a certain number of homes. Each homesite is predesigned for roads, water, sewer, electrical services, stormwater drainage, telephone, sewer service, cable TV, and, in newer subdivisions, fiber-optic Internet access. Every service that is necessary for the modern home is included with the cost of a new subdivision homesite. When you build on lots outside of a subdivision, and on some lots in older subdivisions which were built before utility services were required to be provided, utility services and stormwater drainage plans have to be engineered from scratch. This extra engineering cost will make a home on an offsite lot more expensive to build in most cases.

Before you finalize the purchase of a parcel of land outside a modern existing subdivision, you would be well advised to have your builder look at the property with you. If your builder is familiar with the special requirements of building outside of an engineered subdivision (keep in mind that not all builders have this necessary experience), he will be quick to spot potential problems that could add to the cost of building your new home. In many cases where a lot may be particularly difficult to work with, he may advise you to obtain the services of a professional engineer.

In these cases, a professional engineer (PE) will examine your property and make specific recommendations on the design of your wastewater treatment and sewage systems, and how your lot should be graded and landscaped for efficient stormwater drainage. Unfortunately, most new homebuyers are reluctant to obtain a PE's services because they think he is too expensive. A PE is paid for his time on an hourly basis at a comparable rate with your CPA, lawyer, or any other skilled professional. Even though this rate may seem high on the surface, the potential to help avoid costly or sometimes irreparable mistakes later in the building process makes an engineer one of the best bargains you will ever find. There is absolutely **no doubt** in my mind that many of the mistakes I see with offsite work could be avoided if only a few hundred dollars had been spent on a PE at the offset. As a conscientious builder, I would rather spend $500-$800 on an engineer than create a situation that could cost many thousands of dollars or be impossible to fix later on.

One such impossible problem happened a few years ago in a local high-priced community where each home was required to have well and septic tank installations. This particular home was built by a competing builder before the septic tank or wastewater treatment system had been properly thought out and designed. When the local health department, the governmental agency that regulates septic tank installations, came out and tested the site, they determined that the septic system must be elevated three to four feet above the existing ground. Unfortunately, the finish floor elevation of the home was already placed at the standard eight to twelve inches above the existing ground, rather than being elevated to compensate for the septic mound in the backyard. Because of this mistake, the great room of this stately home with its elegant features and expansive windows will now and forever overlook not the pristine forest and gently flowing stream adjoining the property. Instead, the homeowner and his guests can now enjoy a fantastic view of the chest-high dirt mound that is a permanent fixture of his backyard.

Another instance of an irreparable problem happened in the same area where a similar septic tank mound was installed on a home that had its finish floor elevation set too low. Like most new homebuyers, the homeowner moved in during the dry season of the year when there was absolutely no rainfall and therefore no potential for a drainage problem. When the first rainstorm erupted onto the scene several months later, the septic tank mound acted as a dam and blocked stormwater from draining out of his backyard. As the water level in his backyard rose with each falling raindrop, the homeowner realized he had a problem. Normally when a drainage problem like this comes up, you can simply bring in extra fill dirt and shape the ground around the house so rainwater will drain away from the back yard. Unfortunately, in this case the home was set so low that if the builder brought fill dirt into the backyard, the dirt would have been piled against the homeowners' elegant picture windows. So because the home's elevation was set so low, the homeowner now has to continually live with the reality that his backyard could become a lake every time a storm cloud appears on the horizon.

These are the kinds of situations that would not happen if the services of a PE and an experienced builder were used to ensure that the initial design of the septic system and stormwater drainage system were properly done. And although an extra $500-$800 dollars may seem like a large amount, chances are the homeowners in these examples would gladly pay that much if they could start the building process all over again.

When your builder looks at a site outside an engineered subdivision, the four main things he considers are the following:

1) access to the property from a public road
2) jurisdictional land or wetlands if any
3) individual water supply and wastewater treatment systems
4) drainage

1) Access From A Public Road

Usually forgotten about when property is initially examined, providing access from a public road can lead to many extra expenses if not considered carefully. When an offsite parcel is located directly off a paved road, you will usually only need a typical concrete driveway like you would find in any subdivision to gain access to your home. But unlike a subdivision where the typical driveway is forty to sixty feet long, an offsite driveway may be _several hundred_ feet long. Extra fill dirt and drainage structures or culverts may also be needed to allow ditches that flow underneath the driveway to continue transporting stormwater.

And when a driveway is located off an unpaved road, you will usually have to upgrade it from a simple concrete pad to an actual paved road. Such a road will almost always require some kind of stabilization material such as crushed stone or sand to be added underneath the road to prevent settlement. Also, you may need to use asphalt or paving material to surface your road, which will cost considerably more than the concrete used for a simple concrete driveway. When you take all of these factors into account, it is easy to see how an offsite driveway can end up costing substantially more than a concrete driveway in an engineered subdivision.

2) Wetlands

Also referred to as jurisdictional lands, wetlands are currently under the special protection of many state and federal agencies. Regulations governing wetlands are new compared to other governmental sets of rules, but they _must_ be considered before any work begins if the potential for wetlands exists.

When I first started in the building business, wetlands were easy to determine – if you walked across a piece of property and your shoes got wet, you had wetlands. Nowadays wetlands are defined not by moisture but by the vegetation growing on the land. To define wetland areas, a biologist working in conjunction with an engineer will mark the jurisdictional or wetlands line based on the vegetation she sees. Then she will ask the governmental agencies in charge to approve her interpretation of the jurisdictional line location. When the agencies express their approval, the surveyor will locate the biologist's flags used to mark the line and plot the jurisdictional line on a survey of the property. The wetlands areas beyond these jurisdictional lines cannot be altered or disturbed without a special permit.

The job of delineating wetlands should be handled by a professional engineer working in conjunction with a biologist in **every case without exception**. If an experienced builder suspects that wetlands are somewhere on your property, he will call a PE to investigate prior to doing any building whatsoever. Our industry abounds with horror stories of people who disturbed wetlands and were forced to either pay heavy fines <u>and/or</u> pay to convert wetlands back to their natural state. No experienced, knowledgeable builder would be foolish enough to take a chance when it comes to wetlands and neither should you.

3) Individual water supply (well) & septic tank systems

Individual water supply and septic tank systems are usually regulated and permitted by the local health department. A building permit for a new home will usually not be issued until the health department has given their blessing to the well and septic tank design on your new home.

For an individual water supply system, most local, reputable well-drilling companies can tell you what you will need when you have a well installed including:

- approximately how deep your well must be dug,
- the type of well required, and
- additional equipment needed for your well system such as pumps, aerators, storage tanks, etc.

There are several important and frequently forgotten items you should remember when you are pricing your well water system. The first is the cost of providing electrical service from the house to the well's electric pump. The farther from your home the well is located, the more it will cost to run electrical lines to your well. Second, in areas where the water has a high sulfur content or merely tastes bad, an aerator may be needed to filter your water. This aerator system may require you to buy an additional storage tank to keep a supply of aerated water on hand for immediate use in your home. Finally, when using a well system, you need to carefully plan where you will put it. Local regulations will require the well to be a certain distance from the septic tank and drain field area so that sewage will not leak into the water supply. In addition to your own septic system, it is also important to keep the well a certain distance from other septic systems on neighboring lots. Keeping these things in mind when you are planning where to place your home on your homesite will help ensure that only pure water comes through your faucets.

When I think of all the regulatory functions the health department handles, I suspect that the regulation of septic tank permits would be the one function they would most like to delegate to another agency. Horror stories abound in every community about the excessive and horrendously expensive regulations forced on some poor homeowner by the local health department when he went to apply for his septic tank permit. What these people do not realize is if the health department failed to enforce strict guidelines for septic tank installations, the systems would fail to operate in wet weather. Then with every rainstorm, the homeowners would receive a big surprise as their plumbing system backed up and raw sewage seeped into their bathtubs and stopped up their toilets. The horror stories you hear now would seem tame compared to the substantial health hazards that would occur as disease and foul odors drove homeowners from their homes.

To protect the public from such horrors, the health department has no choice but to enforce very strict and sometimes costly regulations. When your builder applies for a septic tank permit, the health department will inspect the site, determine the type of topsoil existing on the lot, and access that soil's ability to accept and filter wastewater. After this is determined, the health department official will find the elevation of the water table (the top of the saturated soil layer in the ground) and determine the size of the septic tank needed and the area of drain field required for the leaching system.[i]

As with your well water system, when you are finding the cost of a septic tank system on your property, you can call any reputable septic tank installation company in your area and find out what is usually required by the health department. Such a

Diagram of a Typical Septic Tank System

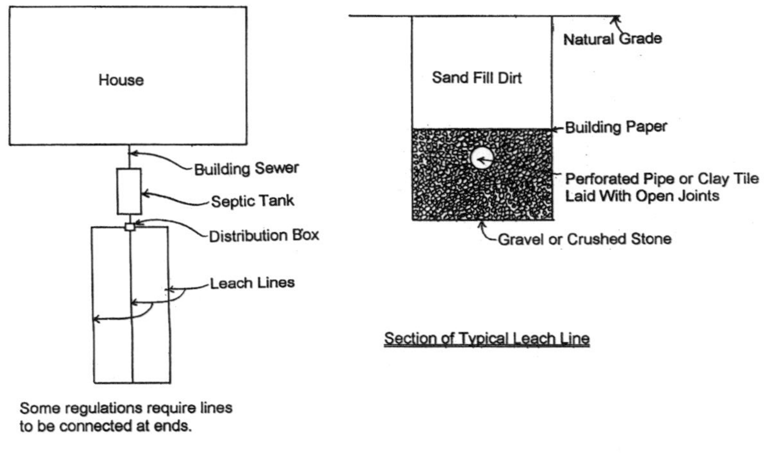

Individual sewage disposal systems are strictly regulated by the local Health Department. Although design and construction practices vary widely, the basic principles of septic tank systems are the same as what we describe here.

Sewage flows from your home through a solid pipe called the building sewer pipe into a watertight box (the septic tank). Solid waste remains in the tank suspended in water until they are broken down or liquefied. The resulting liquid (effluent) flows through a solid pipe into a small, watertight box called a distribution box. From there, the effluent flows into the leach lines and into the ground. Leach lines are perforated pipes or pipes with holes which allow the effluent to leak out into the ground. In order to provide additional openings and facilitate the absorption of effluent into the ground, the trenches surrounding the leach lines are filled with gravel or crushed stone. (see trench diagram).

company should be able to give you an idea of what the price of an ordinary drain field and septic tank system will cost to install.

Unfortunately, once you are given this price, you only have part of the story. Many times I have had clients get a quote from a septic tank installer and become horrified when I tell them the price of their system will be more than they were quoted. You see, when the health department determines the elevation of the water table in the soil, they require that the drain field be a certain distance above that level. So if you have granular, well-draining soil and a water table located far beneath the ground, then you will not need additional fill dirt to raise the drain field above the water table. If, however, you have impervious soil layers such as clay and muck above a water table close to the earth's surface, then you may have to truck in additional fill dirt and raise the elevation of the drain field bed. In some cases, I have seen these beds have to be raised as much as **four feet** above the ground. These raised beds are horrendously expensive. And to add insult to injury, these beds also require you to raise your home with additional dirt so the waste products from your home will drain downhill into the drain field. As if this were not enough, you also have to add fill dirt around the home so stormwater will drain away from your home. All this extra fill dirt can run as much as **five to ten times** the cost of the septic system and drain field alone. Usually when I give my clients this bit of bad news, I have to stand behind them to make sure they do not hit the floor as they pass out from the shock.

Now there is another method of running a septic tank that does not require quite as much fill dirt. Instead of using just a normal septic tank, it has what is called a dozing tank- an ordinary septic tank with a large pump inside of it. This pump is connected to a float inside the tank which will automatically engage the pump when the tank fills up to a certain level. This pump then transports the sewage into the regular septic tank where the force of gravity pushes it downhill to the drain field for treatment.

The reason why I usually avoid this type of septic system is because the system pump requires periodic maintenance. A dozing tank is effective in certain instances where you need to avoid raising the floor level so high. However, like a normal septic system, a dozing tank also requires a large mound to be placed around the septic tank in many cases. Therefore, great care must be exercised to prevent situations like the one we talked about earlier with the mound damming the homeowner's backyard and turning it into a lake. The location of a septic tank mound and the elevation of your home's foundation is critical for determining the pattern of stormwater drainage in wet weather. The layout and elevation of the home, as well as the pattern of stormwater drainage, should be thought out completely and drawn on paper before any building is considered.

4) Drainage

Building officials across the country will tell you that the most common complaint they receive on new home construction is drainage. The reason is because people look at their potential homesites on bright, sunny days when everything is dry and forget about the rainy season that will soon be upon them. Also, most homes are built during dry months, so many drainage problems do not appear until after people have already moved in. Several months go by, the rain sets in, and one morning the homeowner steps into his backyard and finds he has a new swimming pool he did not order. Then the realization dawns that he has a drainage problem, and the local building department becomes flooded with drainage complaints.

I remember one story a fellow builder told me about a couple who had their own lot and paid no attention to drainage considerations whatsoever. The couple's three acre lot backed up onto a creek and after the various governmental agencies were through defining the jurisdictional wetlands and setbacks were taken into account, the couple barely had enough room to build their home. When the builder told the property owners that they may have drainage problems with this large wetland area and strongly suggested that they find another lot, the couple balked and insisted that their dream home be built on that spot. As with most homes, this one was built during the summer season when the wetlands were dry as a bone and there were no drainage problems in sight. When the rains came three months after they moved in, the creek level swelled, and the homeowners found their back yard completely flooded. To their chagrin, when they looked in front of their home, they realized there was less than 100 feet from their home to the street on a **three acre lot!** When they realized they had made a grave error not listening to their builder's advice, they tried to sell their home. Unfortunately for them, the home was nearly impossible to sell because potential buyers would see how high the home was built and the lake in their backyard and run the other way.

During the planning of a subdivision, a site engineer will take factors like these into account when he designs the drainage system. He will make sure that the slope of the ground on all homesites is adequate and set the finish floor elevations on each home sufficiently high for the drainage system to work correctly. On an off-site lot, none of these calculations have been done beforehand. Therefore, your builder may have to bring in some extra fill dirt and put in a little overtime to make the drainage on your homesite work the way it should.

Other Considerations

When building outside of an engineered subdivision, people sometimes are concerned about whether they can get electrical power to their home if they are located far from power lines or generating stations. Luckily for these people, local electric authorities are required by law to run electrical power to your home no matter how far you are from their existing power supply lines. In exchange, the government grants electrical authorities a monopoly on electrical services in a given area. This system was instituted by President Franklin D. Roosevelt during the 1930s when he established the Rural Electric Authority (REA). The purpose of the REA was to provide electrical power to people in rural areas who would otherwise not have service available. Giving electrical authorities a monopoly in their area allowed the running of expensive electrical wires in sparsely populated areas to become a profitable and worthwhile venture where it otherwise would not be. So in order to keep their monopoly, your local electrical authority is required to provide you with electrical power, even if they have to string 100 miles of electrical wiring to get to your front door.

Although building your home outside of a subdivision and establishing yourself as a "gentleman farmer" is more expensive and difficult, it can be one of the most rewarding things you will ever do. During those days when you want to leave the working world behind, having the privacy of your own lot provides you with freedom and peace of mind that is hard to put a price on.

In fact, as soon as I was able to afford it, I sold my house in the suburbs and built my dream home on a beautiful fifteen acre parcel away from the city. When I drove up my driveway and unlocked my front door, I did not want to have to deal with anybody. Establishing myself as a "gentleman farmer" on an undeveloped parcel of land was the best way to achieve this freedom.

But like all historic liberties, your sanctuary from the trials of everyday living will bring a heavy price tag. The extra builder's costs, the extra cost of fill dirt for your septic tank, stormwater drainage engineering, and higher home elevation will most likely make building on your own lot much more costly than building in a subdivision. Whether or not the freedom you will enjoy is worth this extra cost is something no book can tell you. It is a question only you and your family can answer.

Highlights of Chapter 8

- Land outside of a subdivision is usually more expensive, not less, to build on.

- A builder has his operation set up to build multiple homes in a subdivision. Therefore, it costs him more to reset his operation to build a single home on an offsite lot.

- Usually stormwater drainage and utility services must be engineered from scratch on offsite work – making building a home more expensive.

- In some cases, an offsite lot may need the services of a professional engineer (PE). Although his fee may seem high, it is low compared to the mistakes he will help you avoid.

- To gain access to your property, you may need a driveway several hundred feet long. This length makes it more costly to construct than subdivision driveways.

- If there is any danger of wetlands being on your property, you must call a PE to investigate <u>before</u> any work begins.

- Well and septic tank systems are regulated by the local health department which enforces strict and sometimes costly regulations.

- When finding the cost of installing a well, consider: how deep it must be dug, type of well needed, any necessary extra equipment, and cost of providing electric service to your pump.

- Your well must be located a minimum distance from your septic tank and drain fields and neighboring septic systems to avoid drinking water contamination.

- When pricing a septic system, call a reputable company for a quote on a normal septic system and drain field. Then find how much fill dirt will be needed to raise the drain field to the height required by the Health Department.

- A dozing tank is effective when you need to avoid raising the floor level of your home to an excessive height; however, the pump inside of it requires periodic maintenance a normal septic tank does not require.

- Your builder may have to truck in excessive fill dirt to make your drainage work properly since no drainage engineering is usually done beforehand on offsite lots.

- Local electrical companies are required to run electricity to your home no matter where you live. In exchange, the government grants electrical authorities a monopoly on electrical services in a given area.

- The extra builder's costs, extra fill dirt, stormwater drainage engineering, and higher home elevation combine to make building on your own lot more costly than building in a subdivision.

[i] For illustration of septic tank terms, see diagram of septic tank system.

Notes

Notes

Chapter 9:
Designing Your New Home

During my college days, I remember a young architect who taught our basic architectural design course and introduced himself by talking about the first home he ever designed. It seemed that one of his friends asked him to design a long, low, rambling one-story ranch home with one distinguishing characteristic – he just had to have a foyer with an ornate staircase and balcony.

Of course we all laughed about this elegant stairway to nowhere, but the professor had eloquently made the point that sometimes what your client wants is difficult, if not impossible, to include in a new home design. While I was a student, I thought that this made home design frustrating and tedious; however, after I graduated and began working, I discovered that this is actually what makes new home design so interesting. Home design is based on the practical application of natural sciences and mathematics, but it is these streaks of fantasy that provide the creative spice for architectural planning. It is imagination that separates a mere shelter against the elements from a home that is personally tailored to your needs and desires.

When you begin designing your new home, you may want to find something a little more practical and less expensive to use as a conversation piece than an ornate staircase to nowhere. The layout of your home should be based on the practical needs of you and your family, tempered with your own personal flair and eye for design. Most people in the world, including myself when I built my new home, fall into the category of having a finite amount of money to spend. And for those of us whose last names are not Kennedy or Rockefeller or Gates, the thought of hiring an architect to design our dream home has crossed our minds at one time or another. But who wants to pay $150+ an hour?

The average person simply will not hire an architect to design their new home. Instead, the vast majority of homes bought in the United States are builders' model homes which have been modified to suit the needs of a specific client. This is the formula most of our clients have used during the past thirty years, and it has served them well. If you have a limited amount of money you want to spend on your new home, chances are this system will be the most practical.

Pitfalls to Avoid When Viewing Model Homes

When you choose a home from a builder's selection of models, there are several pitfalls you must avoid to prevent making mistakes in your initial selection. The first mistake most people make is they look at how the model home is decorated, instead of concentrating on the home itself. You must remember that once you buy your new home, the model furniture and decorations will leave with the builder, and yours will be put in their place. If your furniture is not as elaborate or pretty as what the interior decorator chose to spruce up the model home, you may find that the house is not as charming once it is yours.

A second big mistake people make is they buy a home simply because the site agent is friendly and amiable. In the midst of getting to know their site agent and becoming their trusted friend, the new homeowners forget that they will be living in this home long after their agent has moved on and forgotten who they are. A friendly smile and a good demeanor will not mean much down the road when their roof begins to leak and they discover that their builder went out of business the year before.

Homeowners also do not realize that their site agent has no way of guaranteeing that their home will be a quality product because the salesmen are not out there everyday overseeing the construction of your home. A site agent can spew endless statistics about how good a builder is and how many awards he has won, but unless your site agent is out there supervising the construction work, he cannot guarantee that your home is built properly. Now do not get me wrong; awards are pretty and they look good hanging on an office wall (I know mine do), but they are no guarantee that a builder knows what he is doing. The only way you can feel confident in knowing that your home will be built well is to make sure you have done your homework and chosen your builder according to the research techniques we talked about earlier, not by taking the word of your friendly neighborhood site agent.

Structural Considerations of Your New Home Design

A new home design is based on the practical needs of you and your family, sprinkled with the spice of imagination. Designing is a very interesting process and a lot of fun for you and your builder/designer. In this process, your builder/designer has two jobs:

1) meet your family's needs and desires, and
2) make sure your new home satisfies the requirements of the local building code.

The building code is set up by local municipalities and lists requirements that have to be met by designers and builders as they go through the design and building process. A building code is actually several codes that cover building items such as: plumbing, mechanical systems, electrical systems, handicap accessibility requirements, energy efficiency, and structural considerations. Structural considerations include: gravity loads, wind loads, snow loads, and even seismic or earthquake loads in some areas.

One part of the building codes that has received tremendous coverage by the press in recent years is a building's ability to resist high winds or "wind loads" in new construction. Wind loads became a major concern after the hurricanes Hugo and Andrew caused so much damage in the southeastern United States. The massive devastation these storms caused quickly focused attention on the need for greater care in the design of buildings to resist high winds. In response, more attention was given to those areas of the building code which govern wind loads and your home's ability to resist them.

The building code contains maps that give the highest wind speeds for each area based on historical weather data for that particular region. These maps tell you that homes built in certain regions of the country need to be able to withstand greater winds. The wind forces pushing against your home during violent thunderstorms increase greatly as you move across the country, especially if you build in coastal areas where hurricane winds off the ocean are much greater than winds in inland areas. If you have ever sailed a boat before, then you realize that the larger your boat's sail, the bigger the area the wind pushes against and the faster your boat will go. The same holds true for your new home. When you have tall, wide walls outside your home, the wind treats them as big sails and pushes against your walls as hard as it can. The bigger your walls, the more the wind can push. When you were out on your sailboat, this wind that pushed you onward was a good thing. But in your new home, the last thing you want is a violent windstorm blowing your home down the street.

To prevent your home from blowing away, wind loads are resisted by "shear walls" or walls that run parallel to the direction of the wind. These shear walls act as braces, reinforcing the walls that are perpendicular to the wind so they do not get knocked down like very large dominoes. Your designer uses formulas from your local building code to calculate how long the shear walls in your new home must be to keep it standing.

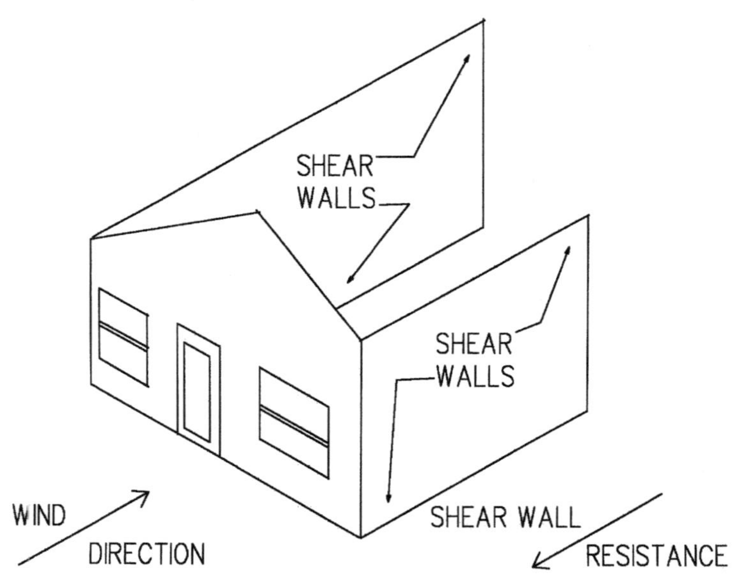

SHEAR WALL SEGMENTS RESISTING WIND FORCES.

Now I know some of you may be asking yourselves, "Why should I care about shear walls? It's my designer/builder's job to make sure I have them, so all I have to do is come up with a design and he will stick them in where I need them, right?" Well, yes and no. You see, a shear wall segment is only allowed to have openings no greater than one foot by one foot. This means that you cannot put a window or a door or an opening of any practical size in a part of a wall used to resist wind loads. In most home designs it is usually not a problem if you want to add a window or door to the sides of the home where you often have large segments of wall space with no windows or doors. However, the problems commonly occur when you want to place extra windows or doors on the front or rear of your home where you already have large openings to maximize your view.

When you have selected a homesite which has a nice view of the beach, woods, lake, mountains, or other picturesque scenery, the first thing you will naturally want to do is put as many large windows as you can in the front and back to maximize your view. When you do this, you will decrease the amount of shear wall available to resist winds that hammer the sides of your home. In order to make up for the lack of front and rear shear walls, the builder will have to place some extra hardware such as metal straps or extra wall and roof reinforcements into the structure. All this extra hardware and engineering will translate into extra cost for you when your new home is built. In some cases, a lot of extra cost. In fact, I have had cases where it cost four or five times as much to add a window to the back of a house as it did to add a window to the side of a home where I had large segments of shear walls.

When you are designing your new home, the main thing to remember is to **stay flexible**. Be aware that minor changes in your design can lead to big savings or big costs, depending on what you do. For example, if you make a room wider, the structural members of the roof and the floor will have a greater distance to cover between their supports. This means that to support the weight of your new home, these members will either have to be made larger or have to be spaced closer together, both of which will mean increased cost. In addition, the more windows and doors you add to the plan, the more it will cost because an opening is more expensive to build than the section of wall it replaces.

A trained, experienced designer/builder understands concepts like these and will incorporate them into your new home design. Your job is to work together with your designer and to make sure he knows what you want and need throughout the design process. This way he can give you what you want and provide you with alternatives which may work as well as your original thoughts and save you money at the same time. If you remain flexible and open to new ideas throughout the design process, you can get exactly what you want and pay for your new home without breaking the bank.

Determining Your Design Needs

Designing your new home consists of nothing more than narrowing down the many choices you have available until you arrive at your final design. Designing should always start with a floor plan. The floor plan is known as the "workhorse" drawing in a set of blueprints because it is the drawing most used during both the design and construction phases of your home. From the floor plan your designer will come up with a complete set of blueprints like we saw in Chapter 2. This complete set of blueprints includes:

- the foundation plan,
- exterior elevations,
- interior elevations or details,
- electrical plan,
- site or plot plan,
- roof framing plan, and
- various architectural details and sections.

After you have the floor plan, coming up with the other blueprint drawings can be considered nothing more than "fine-tuning" your plan until your design is completely finished. Some experts feel that there are only so many basic floor layouts in existence and that all other floor plans are merely modifications of these. When you start comparing floor plans from different area builders, you will probably agree.

When you are deciding what floor plan you would like to begin with, you need to first narrow down your possibilities and determine exactly what you need in your new home. In the following several pages, we have assembled a concise but informative list of items to help you discover what you are looking for. As you look at each item, do not just think merely of what your family needs in a home right now, but what they will also need in the future – five, ten, or even twenty years down the road when your needs could be much different than they are today.

Number of Floors

When you are deciding the number of floors your home should have, there are several questions you need to ask before committing to a one, two, or three story home with a basement. These questions include:

➤ **Do you want to climb stairs?**

➤ **If yes, will you want to climb those same stairs in five or ten years?**

➤ **Do you want to lug things such as furniture and clothes upstairs?**

➤ **If you want stairs, do you want the master bedroom up or downstairs?**

➤ **Do you want the children's bedroom(s) up or downstairs?**

➤ **Do you want your children's bedroom(s) next to your master bedroom or away from your bedroom?**

➤ **Will you have elderly relatives coming to live with you?**

➤ **If yes, will you need a separate master bedroom for them?**
With the aging population, adding a separate bathroom to a bedroom and making a home with two master suites is a practical and attractive option for children taking care of their aging parents.

➤ **If yes, you will need to make sure that elderly relatives have their bedroom on the ground floor**.
This way, they do not have to negotiate stairs that they might trip over and hurt themselves.

Number of Bedrooms

➤ **Every home needs at least three bedrooms.**
Over the years, we have discovered that having less than three bedrooms hurts the resale value of your single family home. Remember, you can always use three bedrooms if you need two, but your future buyers cannot use two bedrooms when they need three.

> **Do you need to give your children each their own bedrooms?**

> **Will you have other family members coming to live with you for a considerable length of time?**
> If so, you will need to have a separate bedroom for them – and it would be a **very** good idea if they had their own bathroom as well.

> **Do you want to have an office at home?**
> An additional bedroom is perfect to modify for a home office. Just remember to add plenty of electrical outlets, a dedicated electrical circuit for your computer, and a separate phone line for your computer modem and/or fax machine.

Number of Bathrooms

The number and types of bathrooms your home has depends on any number of things:

> personal tastes,
> home budget,
> layout of your home, and
> how much you entertain.

Bathroom designs vary greatly from person to person – from a bath with only a toilet and vanity, to elaborate master baths with showers, double vanities, whirlpool tubs, and any number of custom items. The number and style of bathrooms in your home is dictated by personal preference and budget more than any other room in the home. So as you weigh design options for the most personal and private rooms of your home, keep these questions in mind:

> **Do you want to have access to your bathroom from a hallway, a bedroom, or both?**

> **Should each bedroom have its own bathroom or should the spare bedrooms share a bathroom with guests by using a hallway access door?**

> **Should you add a conveniently located bath for guests?**

- ➤ **Should you add a half bath or full bath near the garage so that you can wash up after working in the yard <u>before</u> you track dirt through your home?**

- ➤ **Should your bathroom have a separate room or compartment for the toilet?**

- ➤ **Should your bathroom have a shower, a tub, or both?**

- ➤ **Would you like to add a whirlpool tub to your bathroom?**

- ➤ **Would you like to have separate his and hers vanities in your bathroom, or are you content to share one?**

- ➤ **Would you like to access your closet through your bathroom or only from your bedroom?**

- ➤ **Do you think you may add a pool or hot tub to your home?**
  If so, you may also want to design a bathroom(s) with access to your backyard or back porch so you can go straight into the bathroom without dripping water through your home.

Fundamental Design Questions

After you have determined the number of floors, bedrooms, and bathrooms you want in your new home, several basic design questions still have to be answered before you can start picking out and designing your new home.   Here are a few of the fundamental questions that will allow you to narrow down your search for the perfectly designed home.

- ➤ **Do you want a great room or a family room with a separate formal living room?**

- ➤ **Do you want a formal dining room?**

- ➤ **Do you want all of your bedrooms together or do you want the master split from the rest of your bedrooms so you can enjoy more privacy?**

- ➤ **Would you like a vaulted ceiling, an eight foot, a ten foot, or twelve foot flat ceiling or higher flat ceiling in your home?**

> **Would you like these ceilings to be different heights and styles in different areas of your interior or would you like them to be carried through the entire home?**
Remember, the higher the ceiling, the higher the cost, so make this decision carefully.

> **Specialized architectural details,** such as tray ceilings, where the ceiling is indented are a favorite option among higher-priced homes. If you are looking for upgrades, such architectural details as arched openings and special molding packages may be something to consider for your new home. Just remember to balance your taste for extras with what you have already budgeted for your new home.

> **What would you like the pitch of your roof to be?**
High pitched roofs were originally designed for northern climates so that the snow would roll off the roof easily, instead of piling on the top of the house and causing a structural collapse. However, many people today see high-pitched roofs as status symbols and build them in climates like here in Florida where they serve no practical purpose and although they look nice, do increase the cost of roofing your new home.

> **What kind of roofing material would you like to use?**
Basic asphalt shingles are the standard with a fifteen or thirty year warranty, but architectural shingles last for thirty to forty years if you are willing to pay a little more money up front. Aluminum roofing systems have also become popular in the last few years, but clay tile roofs are the rage in southern Florida and the Southwest region of the country. And with manufacturers inventing new products all the time, you have an array of quality choices, all of which change the character, price, and style of your home significantly.

> **What kind and size garage would you like?**
Do you need a one, two, three, or even a four car garage or do you live in an area where you do not need a car? Do you like a front entry garage, or would you like a sweep entry garage where you cannot see your garage door from the front of the home? Also, if you have enough room on your lot, you could

include a garage completely detached from your home. If you have a detached garage, a covered walkway might prove to be a beneficial add-on for rainstorms.

➤ **What kind of finish flooring would you like to use in the rooms of your home?**
Would you like carpet, vinyl flooring, tile, wood flooring, or some other finish flooring for your home? The types of finish flooring and the subcategories of each that you can choose from could fill a book in themselves and range from very inexpensive to the category of "If you have to ask, you obviously cannot afford it." Also, new products and grades of material are coming out daily, making your choice of finish flooring a daunting task. So when you are deciding on flooring for your home, take your time and choose carefully. Each type of flooring can change the character of your home and make or break the budget you have set for yourself.

Kitchens

There are probably more ways to design a kitchen and more variations on kitchen designs than any other part of a new home. A kitchen has to be as versatile as the cooking and eating habits of the people who use it. When you are trying to decide what you need from your new kitchen, here are a few questions to keep in mind that may help your planning go smoother:

➤ **How much open space do you need in your kitchen?**
This depends on several factors such as how many people will be cooking at one time, will you have a lot of people in the kitchen with you while you are cooking, and how much walking will you do between the pantry and the stove or oven.

➤ **Do you want a breakfast area in your kitchen?**
This will depend on whether you want to use your formal dining room or whether you would like to use a more informal setting for everyday meals. Also, some of our clients have even had a breakfast area instead of a formal dining room simply because they found it to be more of a convenience.

> **Do you want an island in your kitchen?**

Some people find that countertops in the middle of their kitchen inhibit the flow of people walking too much, while others enjoy the extra counter space. Also, a number of homeowners place kitchen fixtures such as the range and the sink in their island to make travelling between fixtures easier. Although island cabinets are usually more expensive than peninsula cabinets, because of the additional plumbing and electrical fixtures that are usually included with islands, there are those who think the extra expense is well worth it.

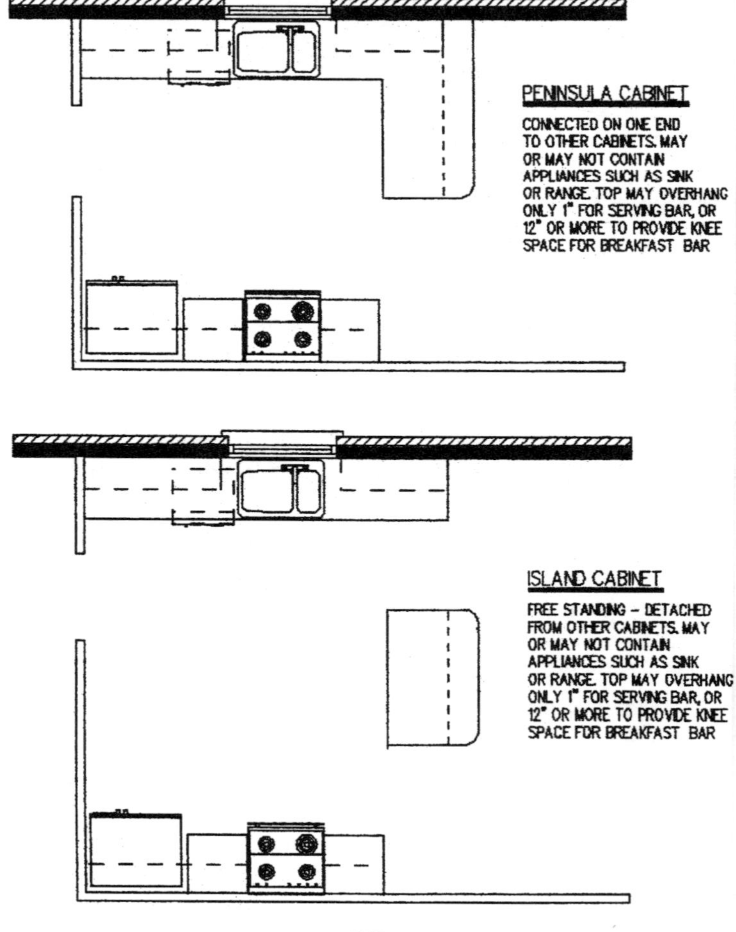

> **Do you want to include a bar in your kitchen?**
This is purely a matter of personal taste, but some like it for the extra counter space it offers, while others like it for its original purpose – the storage and serving of alcoholic beverages.

> **How much cabinet space would you like?**
This, of course, will depend on the number of items you need to store in your kitchen and the amount of space you have in your kitchen to put cabinets. The more you have to store, the more space you will need.

> **Would you like to include a microwave over your cooktop or range?**

> **Would you like a built-in range with an oven underneath or a cooktop?**
The built-in range allows you to have an oven with a minimum of space, while a cooktop allows you more layout and design freedom with your kitchen due to the fact that it is smaller than the combined unit. Just remember, if you do use a cooktop instead of a built-in range, you will need to design an extra space in your kitchen for a built-in oven.

> **How much counter space do you want in your kitchen?**
Again this depends on how much space you need and how often you are going to cook. If you cook elaborate meals often, you will probably need more space than someone who cooks TV dinners most of their life, unless, of course, they cook quite a few at each sitting.

> **What kind of lighting would you like to include in your kitchen?**
The variety of lighting you can have in your kitchen is endless; however, there are some standards that are used in most kitchens today that work well. You can have fluorescent lighting which gives you a lot of light with low energy costs or track lighting for a "warmer" kind of more intense light if energy savings are less important to you. You are not limited to these standards when you pick your kitchen lighting, but these have proven to be the most versatile for our clients.

> ➤ **Do you want your kitchen facing the front or the back yard?**
When your kitchen faces the front of the house, you are able to see people coming to your front door while you cook or wash your dishes. If your kitchen faces the back yard, you can watch your children play or swim in the back. Very few people have requested a kitchen in the middle of the home, because most either want to watch the front or back yard while they are cooking.

I Know What I Want – Now What?

When you have decided what you want in your home, you are ready to collect floor plans from various builders' models and pick out those that meet the criteria you have established after going through the checklist above and have a basic layout that appeals to you. When you look at these floor plans initially, **DO NOT** look at the size of the room or the overall square footage. Instead, look at how your family and guests will travel throughout the house or "traffic patterns" and see if they feel comfortable. After you narrow your plan choices down to three or four favorites, check back with these builders and find out how much these plans cost to build. These model prices will give you a good idea how much you will need to purchase your new home. Since you have already gone through the mortgage prequalifying process or will go through it shortly, you know exactly how much money you have available. Now the only trick is discovering how much you can modify your new home design and stay within this budget.

There is an age old adage in architectural design that states: "Form Follows Function." This means that a fire station should look like a fire station, a grocery store like a grocery store, and so on. Architectural design also has a lesser known rule which dictates: "Form Follows Money." As a designer, if I come up with an impractical building design that no one can afford, then it will never be built no matter how beautiful it looks on paper. I remember a perfect example of this from my college design course when the professor talked about how it is possible to build a mile-high skyscraper. The reason it has never been built is not because man does not have the technology to build it, but rather no one could justify the cost of building it. Another example occurred in India with one of my favorite buildings, the Taj Majal. This church was built in the 17th century A.D. and encompasses acres of interior space. The reason there is only one building of its kind is not because we cannot duplicate it with modern building practices; rather, no one can come up with a practical reason to foot the bill.

As with these examples, your home will become too expensive and impractical to build if you do not keep one eye on your budget during the design process. You need to realize that anything you do, whether it is adding a $5.00 potted plant or a $20,000 marble floor package, costs money – money that can add up quickly. In order to keep tabs on the cost of your extras, make a list of the items you want to include in your home and put the price of each one beside it. That way, you can periodically sum up the amount of your extras and make sure you have not gone over what you can afford.

At this point, you have narrowed down your choice of floor plans to three or four candidates and you have prices on these models from local builders. Now it is time to look at the individual room sizes and see which ones need to be adjusted. A lot of people get concerned when we begin to focus on room sizes because they have a hard time visualizing the size of a room by a written dimension. For example, when I say a living room is twelve-feet by fourteen-feet (12'x14'), most people cannot visualize this room. An easy way to get over this problem is go back to the home or apartment you are living in right now, measure the room sizes, and compare those room sizes with what is on your plans. For instance, if a house plan has a secondary bedroom that is ten-feet eight inches by eleven feet (10'-8"x11' or 10'-8"x11'-0"), and you measure your bedroom and find it is eleven-feet by eleven-feet (11'-0"x11'-0" or 11'x11'), realize that you are pretty close to what is on your floorplan and decide whether you need to increase the size of that room. After the bedrooms, go through the same process with your family room, kitchen, dining room, and all of the other rooms in your home and see where you need more space and what can be left alone. During those rare occasions where you find you can actually _decrease_ the size of a room, you can often take the extra footage and use it someplace that needs a little expanding.

During the design process, it is important to discuss with your designer which rooms you want to increase or decrease and let him review how the plan can be adjusted with you. With some layouts, it is more economical to change room dimensions in one direction than another for structural reasons. A good builder/designer will be aware of which ways are the most economical and will help make your changes with the least amount of additional cost.

At this time I need to point out a fact that startles most of my clients. Believe it or not, it is much easier to take a floor plan of a small home and blow it up than take a big home and shrink it to the square footage you want. The reason for this is simple. When you work with a floor plan 200-400 square feet (SF) smaller than the home you want to build, you have the luxury of putting extra square footage wherever you want and improving the plan with each additional square foot. In a larger home the exact opposite is true– any reduction in square footage will make the home worse, not better. Many times I have seen disastrous results when someone would try to shrink a larger home. Cutting out space in a large home distorts room sizes and throws the home's layout completely out of proportion. What had been a nice great room or master bedroom in the large home becomes cramped, claustrophobic, and far less usable in the scaled down version. People are reluctant to look at a smaller home when they are looking for something larger because they are so stubbornly hung up on the amount of square footage their home has to have. But expanding a smaller home gives you the freedom to tailor your home to you by adding square footage exactly where you want it. And isn't the freedom to make your home exactly the way you want it the reason you are building a custom home in the first place?

I remember a home we built several years ago where we took one of our model homes that was 1,771 square feet (SF) and increased the size of the home to more than 2,700 SF. Our clients were ecstatic when they began using their design freedom to increase their cabinet space in the kitchen, increase their breakfast area, upgrade their cabinets and carpet, and increase their living room from 12'x15' to 21'x28' so they could entertain their guests or whatever football team happened to pop in on them unexpectedly. By increasing the size and options available in that particular model, our clients took a good plan and made it even better. With the right small floor plan and your own ideas, you can make your good design even better too.

After you have taken a little time and gone through this intricate process, you should have a working floor plan with the number of rooms you want, traffic patterns[i] you are comfortable with, and basic room sizes that meet your needs. Unfortunately, too many people worry about overall square footage at this point when they should be **concentrating on the size of each individual room** in their home. If the size of each room is adequate and you have comfortable traffic patterns with a good basic home layout, then the **overall square footage will take care of itself**. The fact that your home has a little more or a little less square footage than someone else's is nothing to worry about. It is the amount of space you can actually use in your home that counts.

A perfect example of this happened several years ago when one of my clients brought in a plan they had drawn themselves and asked me to build it. Their plan had about 2,500+ SF, but when I started going through it, I discovered that some of our stock models with less than 2,000 SF actually had more usable space because of how the rooms were laid out. The reason for this was because their plan had a tremendous amount of hallway space that cut their home to pieces. These long hallways took up living space that had to be heated and cooled, and none of this space was in a room or a closet where it could be used constructively. When I pointed out this wasted space, the homebuyers became indignant and told me they were quite happy with their home the way it was and did not want to change anything. So they wound up paying for 2,500+ SF of home and received less usable space than other clients who had paid for 2,000 SF simply because they did not think about their home's layout carefully enough and would not take their builder's advice.

As you have probably noticed, we have not said much about the exterior elevations and the overall appearance of your new home. This is because we will design our elevations around the floor plan; therefore, we need a working floor plan before we begin to tinker with the exterior. As one of my college design professors used to say, "Now once we have the floor plans and the room sizes, we can then put the architecture on it." Comments such as these make art historians, artists, and even some architects grimace with pain. This is because like most people, they have visions of taking a New England or Tudor or some other traditional architectural style and working their floor plan around this type of façade. This way they feel they can honor a particular style of architecture by showcasing it with their home.

Preserving a piece of our architectural heritage with your home's style is a fine idea but there may be one small problem. Many times it simply is not practical. Most classic architectural styles were created many years ago to satisfy a particular need in a specific geographic area. For example, the designers of old used local building materials that were abundant and therefore inexpensive in their area, but may be very expensive to import into yours. The stone walls and fences from rocky New England fields are predominant in that area, but to import this material to sandy Florida beach homesites would be horrendously expensive. Also, most classic architectural-style homes were built before modern conveniences such as indoor plumbing, electricity, and heating and air conditioning. They contained no provisions for these modern fixtures and would have to be modified in some way to include these fixtures if you wanted to recreate them. Even if today's building codes did not require these conveniences to be included with your new home, it is unlikely that you would want to omit indoor plumbing or air conditioning in order to construct your home in a "true" architectural style.

So when we have a workable floor plan with adjusted room sizes, an adjusted price, and our list of priced extras, it is time to look at the elevations of the builders' models you collected. If you find one or two of these elevations you particularly like, they can be redrawn with your adjusted room sizes and some additional features until you come up with something that looks great. On the other hand, you may discover that you do not like any of the elevations you have collected. In that case it is time to create a façade from scratch. Many times people have brought us pictures from a magazine, photos of existing homes, or pictures of homes they have lived in before and asked us for an elevation with this certain kind of look. As your designer, we use elements from these renderings and generate preliminary sketches for you to review. Then we modify these sketches to suit your tastes and give your home the look you are trying to emulate.

Remember as you modify your elevations and add architectural details and changes to your design, you are also adding to the cost of your home. As we have said before and will say again, it is important to keep one eye on your budget as you make changes to your home. Frequently we will generate two or three alternate facades for a home and let our clients decide which elements they like from each elevation. Often they will say, "I like this detail on this elevation and the way the roof looks on the second one," and so on. We then take the elements they like from each façade and come up with a final elevation for their home.

Obviously when you make changes to your elevation you may also change your floor plan. For example, if you add a bay window to your dining room, you change not only the exterior, but also the square footage and cost of your floor plan. As we fine-tune your elevation, we also fine-tune your floor plan until we arrive at a final, overall home plan that fits your needs and desires.

Once you have the elevation you want, you need to step inside your home and fine-tune your floor plan room by room. At this point you already have the rooms as big as you want them, so now all you need to do is think of how you want each room to look. Hopefully you have had a chance to go back through the builder's model you based your home on and remember how it was done. Now when you combine that basic model with the changes you have made on your own home, you can go through your plan room by room and create the finishing touches that will give your house the comfortable feel of home.

The changes you will consider for each room will include whether you want the finish flooring to be tile, carpet, sheet vinyl, or hardwood flooring, the style and type of windows in the home, and any upgrades to the molding and trim packages of the home, just to name a few. But just as with every other design change, each time you modify a finishing touch in your home you increase the overall cost. Therefore, you still need to keep a running total of how much each change costs because every improvement affects the affordability of your new home.

Some rooms in your home will take more time to develop than others. This will be particularly true when it comes to your dining room or family room and you try to decide what kind of moldings and trim packages would look best. Each room can be as informal or as stately as you choose. With each new molding package or type of ceiling or expensive type of flooring, you can achieve a new level of opulence to dazzle your visitors. You just need to decide now whether that marble fireplace is worth the extra cost. When all is said and done, you may find the satisfaction you get by impressing your in-laws with majestic surroundings is not worth the price you have to pay.

The rooms which will take the most time to develop will be the kitchen and the master bathroom. There are probably more design books available for these rooms than for all the other areas of your home combined. In addition, manufacturers of plumbing fixtures, electrical appliances, and cabinets also publish catalogs showing their products in magnificent settings which can give you design and decorating ideas for your custom home. Again, just remember to keep your eye on cost as well as appearances because these catalogs and design books concentrate on overall effect and not on the cost of their surroundings. After all, they are not actually going to build the magnificent settings they display in their magazines, you are.

While you design your kitchen, you will find that some cabinet layouts may require more space than you think. As a designer I can tell you that one of the hardest rooms to design is the kitchen. It takes a lot of thought and planning to lay out a kitchen that is both easy to work in and affordably priced. Too often I have worked several hours on a kitchen design just to realize that it is too expensive or will not fit in my "perfect" home design. So if you liked the kitchen layout in your original builder model, you might think in terms of keeping that basic kitchen and merely adding cabinet space or shelving or switching a couple of cabinets around to create your own custom space. Most model home kitchens are an excellent balance between desirability and affordability and can be made into exactly what you want with minor modifications. And believe me, a few custom touches here and there are much easier and more affordable than designing your kitchen from scratch.

Just as a model home kitchen may be worth keeping, a model home master bath is also worth a second look. Master baths are also a great compromise between beauty and price and can be slightly modified to serve your needs. Adding a stand-alone shower, a whirlpool tub, separate closets, or a glass door leading into a garden may require only a minor adjustment in the layout of your master bathroom. And making something work with minimal effort will cost considerably less than starting from scratch.

At the end of this extensive design process, you should have a good idea of your final floor plan, exterior elevations, individual room sizes, cabinet elevations, and a list of extras you want added to your home along with their cost. Now it is time for you and your builder to come up with a final price for your entire new home package. If everything turns out perfectly, your final price will be within your budget. You can then skip the next chapter, break ground, and begin building. If everything turns out not so perfectly, (and yes, odds are this will unfortunately happen to you), you will need to read the next chapter **<u>carefully</u>** to ensure that your dream home does not become a nightmare.

Highlights of Chapter 9

- Your builder/designer has two jobs: make sure your home meets your families needs and desires, and make sure your home meets local building code requirements.

- The building code is set up by local municipalities and lists requirements that must be met by designers and builders as they go through the building process.

- To prevent your home from blowing away in a storm, wind loads are resisted by "shear walls", or walls that run parallel to the direction of the wind and act as braces, reinforcing your walls so they do not get knocked down.

- When designing your new home, you must stay flexible and be aware that minor changes in your design can lead to big savings or big costs.

- Your job as a new homebuyer is to work together with your designer and make sure he knows what you want and need throughout the design process.

- Designing your new home is nothing more than narrowing down your many available choices until you arrive at your final design.

- Designing should start with the floor plan. After you have the floor plan, coming up with the other blueprint drawings can be considered nothing more than fine-tuning your design.

- When designing your floor plan, determine exactly what you need in your new home. Think of your family's needs today and also what they will need several years in the future, when needs could be different than the present.

- When you have decided what you want in your home, collect floor plans from various builders and pick out those that meet your criteria.

- When looking at various floor plans, do not look at room sizes or overall square footage. Instead, look at the homes traffic patterns and see if you and your family will feel comfortable living there.

- After narrowing down your floor plan choices, check back with the builders and see how much these plans will cost to build. These prices will give you a good idea how much you will need to spend on your new home.

- Remember as you design your home that "form follows money." Your home will become to impractical or expensive to build if you do not keep an eye on your budget during the design process.

- With your final selection of floor plans, look at the individual room sizes and see which ones need to be adjusted. To get an idea on how big a room is, measure the rooms in your current home and compare them to the room sizes on your home plans.

- During the design process, discuss with your designer which rooms you want increased or decreased. With some layouts, it is more economical to increase room sizes in one direction than another for structural reasons.

- It is much easier to take a smaller home and add square footage to get the home you want than taking a larger home and cutting square footage. Cutting footage out of a larger home ruins the home's proportions and makes a nice home into cramped, claustrophobic, less usable house.

- During the design phase, concentrate on individual room sizes, not the overall square footage of your home. If the size of each room is adequate and the traffic patterns and home layout are good, the overall square footage will take care of itself.

- A home plan is designed around the floor plan. Therefore, the floor plan must be done before the exterior of your home is ever considered.

- Building your home in a classic architectural style may not be practical because of expensive, imported materials or because a certain style did not allow for modern conveniences like indoor plumbing or air conditioning.

- When designing your elevation, look at your builders' models and think about combining elements you like into your elevation. If this does not work, you may have to create a façade from scratch.

- As you modify your elevation, you also add to the cost of your home.

- When you make changes to your elevation, you may also be making changes to your floor plan.

- After your elevation is complete, you need to step inside your floor plan and fine-tune each room by choosing finishing touches such as carpeting and wood trim.

- Your kitchen and master bathroom will take more time to develop than other rooms.

- It may be worthwhile to consider keeping the kitchen and master bath from your builders' model and modifying them to create your own custom space with minimal effort and cost.

[i] **Traffic Patterns**– the most heavily traveled paths people will take when they walk through your new home. These patterns will affect furniture placement and the convenience of moving around each room and going from one room to another in your new home.

Notes

Chapter 10:
Setting A Price/ Redesigning Your Dream Home

Every now and then I have this recurring dream where a client walks into my office and says: "I want a house that makes me feel at home the instant I pull into my driveway. I don't care how much it costs, just design something I want to come home to. I'll pay you whatever it takes – just make it happen."

When work gets slow and the walls start looking a little too familiar, I find myself watching the front door and hoping my dream client will walk through and allow me to design the home I have always wanted to build– a home without limits. Like all designers and builders, I have always dreamed of meeting this person who will pay whatever it takes to get the perfect home. But in more than thirty years I have never even come close to meeting this person, and the odds are that I never will.

Most likely, those of you reading this book are not my dream client because you have a very definite amount you want to spend on your home. This is why the moment when I give my clients the final price on their dream home, with all of the extras they wanted, is one of the most dramatic moments in my professional life.

Once we go through the design phase of a new home, make all the necessary modifications, draw the preliminary sketches, and estimate the final sales price, then comes the moment of truth when the dreaded sales price must be delivered to the unsuspecting homebuyer. And as soon as the magic number leaves my lips, the real estate agent and I take a collective deep breath and go silent, awaiting their reaction.

Over the years I have seen many reactions to this climactic moment. Some laugh, some cry, and one woman even passed out on me for about five minutes. This reaction used to really bother me because I never understood how anyone could get so distraught over a sales price that directly resulted from their own choices. But over the years, my naiveté eroded and I realized that during the design process, my clients' emotions overrode their normally logical thought processes and their wish list became extraordinarily long. And with this long wish list, the price of their home would inevitably go over budget, sometimes way over budget. The reason for this is very simple. When you add $100 in extras here and a $500 option there, these costs do not seem expensive by themselves. But when they are added together, you wind up with a very large price.

During the design process, adding options and going over budget is a very common thing. In fact, this is not only common in residential homes, but also in government projects, roadwork, dam construction, commercial projects, you name it. This phenomenon has happened so often that people have actually coined a phrase for it – "cost overrun."

So if you find your wish list causes a "cost overrun" with your budget, please realize that it is not the end of the world. You are not the first person to ever run over budget, and you will certainly not be the last. All you need to do is sit back, take a deep breath, and examine the changes you have made in your new home. Look through the cost of each of those changes and decide what you absolutely need and what you can live without. This is the time in the homebuilding process when you need to inject as much logic as possible and find a reasonable compromise between what you want and what your budget can handle.

As you did when you first designed your new home, you need to concentrate on your layout, traffic patterns, and room sizes. Ask yourself questions such as:

➤ Can we choose a less expensive homesite before we begin building our home?

➤ Do we really need that third bathroom?

➤ Can we get along with a little less master bedroom, kitchen, dining room, etc?

➤ Would it be cheaper to build a single story instead of a 2-story home?

A major area where you can save money is cutting back on the finishing touches in your home initially and adding what you truly want later on, after the kids have gone to college or money is not quite as tight as it is now. Some finishing touches would include the following:

➤ Cutting back on extra trim molding such as decorative crown molding around the ceiling, or chair railing around the middle of the wall in the dining and formal areas.

➤ Using a less expensive carpet and underlayment pad when your home is built and replacing it with a more expensive upgrade in several years when the kids are not around to destroy it.

➤ Eliminating some appliances such as built-in microwaves and ovens and adding them later or using less expensive appliances.

➤ Installing a fairly inexpensive grade of vinyl flooring that you can replace with ceramic tile later on.

➤ Purchasing less expensive light fixtures and ceiling fans you can easily upgrade later.

➤ Waiting until later to cover and/or screen your back patio.

➤ Placing a minimum of landscaping around your home and adding plants and decorative statuary later.

When your builder determines a price to build your new home, the price should include a total of all costs, big and small. As you remember from our construction estimate in Chapter Two – The Bid Process, all costs mean **all costs**, whether it is a $6,000 truss roof package or a $100.00 ceiling fan. An accurate sales price must include everything, down to the smallest detail.

When you include all of your costs and calculate the cost of your home, trust the numbers you and your builder have calculated. If you have a cost overrun, do not make the classic mistake of thinking you will be able to cut costs during the construction process because **IT WILL NOT HAPPEN.** Construction prices may go up, but they will **NEVER** come down. Your only solution for getting back on budget in a case like this is to redesign and cut back on options until you get a price you can afford.

Remember, it is not unusual to have to redesign your home as many as **five** different times until you get the home you want and can afford. I have had clients go through the redesign process time and time again for weeks on end until they got what they wanted. Now they are living happily in their dream homes. With a little bit of logic and perseverance, you can too.

Over the years I have seen people get discouraged and hurt themselves by not thinking this process through completely. When I started building in 1965, the average home built in a subdivision in the United States cost $17,500. In 1998, the average home cost $181,900[i]. As you can tell from our table of FHA/VA mortgage rate variations at the end of the chapter, the main reason for this drastic price increase is the fact that from the late 1960s through the 1980s, we had a high inflation rate and costs were increasing faster than they had in preceding years.

I remember designing a custom home for a family in 1966 during this period of high inflation. When I gave them the price for building their new home, they reacted with shock and indignation. I can still see the wife leaning over the table shrieking, "These prices are absolutely incredible! They just have to go back down, and we have decided to wait until they do!" Being a new builder in the middle of a recession with a family to support, I really needed the work so I tried my hardest to get her to reconsider. However, no amount of persuasion could sway her from the belief that as surely as new home prices had risen to unparalleled heights in the past few years, they would have to eventually come back down to reasonable levels.

Since then, I have often wondered if this family is still waiting to build their custom home. If they are, then they have seen their ridiculously high price from 1966 of $15,500 "decrease" to somewhere around $135,000 on today's market. And if they continue to wait for price reductions, they can expect their new home to "decrease" to around $250,000 over the next five years.

Cases like this illustrate how waiting to buy a custom home will make the problem of affordability worse, not better. This is because the affordability gap, the gap between average income and average new home prices, has been widening for thirty years and shows no sign of narrowing. Although politicians have been telling the country for years that inflation has been brought to an "acceptable" level, prices have continued to go up. When you build your custom home today, you lock in your sales price at today's rate and pay for your home with tomorrow's dollars. The rising price of new homes then starts working for you instead of against you, because a few years down the road you will find your home is worth several thousand dollars more than you paid for it.

So building a new home now is to your advantage, even if you think you cannot afford it. Although it may be tight on the family budget for the next several years, raises and cost of living increases from your job coupled with rising real estate values will make building your dream home a desirable and practical investment.

FHA/VA MORTGAGE RATE VARIATIONS

EFFECTIVE DATE	PERCENT	EFFECTIVE DATE	PERCENT
April 24, 1950	4.25	March 09, 1981	14.00
April 02, 1953	4.50	April 13, 1981	14.50
December 03,1956	5.00	May 08, 1981	15.50
August 05,1957	5.25	August 17, 1981	16.50
September 23, 1959	5.75	September 14, 1981	17.50
February 02, 1961	5.50	October 12,1981	16.50
May 29, 1961	5.25	November 16, 1981	15.50
February 07, 1966	5.50	January 25, 1982	16.50
April 11, 1966	5.75	March 02,1982	15.50
October 03, 1966	6.00	August 09, 1982	15.00
May 07, 1968	6.75	August 24, 1982	14.00
January 24, 1969	7.50	September 24, 1982	13.50
January 05, 1970	8.50	October 13, 1982	12.50
December 02, 1970	8.00	November 15, 1982	12.00
January 13, 1971	7.50	May 09, 1983	11.50
February 18, 1971	7.00	June 08, 1983	12.00
August 10, 1973	7.75	July 11, 1983	12.50
August 25, 1973	8.50	August 01, 1983	13.50
January 22, 1974	8.25	August 23, 1983	13.00
April 15, 1974	8.50	November 01, 1983	12.50
May 13, 1974	8.75	March 21, 1984	13.00
July 05, 1974	9.00	May 08, 1984	13.50
August 14, 1974	9.50	May 29, 1984	14.00
November 24, 1974	9.00	August 13, 1984	13.50
January 21, 1975	8.50	October 22, 1984	13.00
March 03, 1975	8.00	November 21, 1984	12.50
April 28, 1975	8.50	March 25, 1985	13.00
September 03, 1975	9.00	April 19, 1985	12.50
January 05, 1976	8.75	May 21, 1985	12.00
March 30, 1976	8.50	June 05, 1985	11.50
October 18, 1976	8.00	November 20, 1985	11.00
May 31, 1977	8.50	December 13, 1985	10.50
February 28, 1978	8.75	March 03,1986	9.50
May 23, 1978	9.00	November 24, 1986	9.00
June 29, 1978	9.50	January 19, 1987	8.50
April 23, 1979	10.00	April 13, 1987	9.50
September 26, 1979	10.50	May 11, 1987	10.00
October 26, 1979	11.50	September 08, 1987	10.50
February 11, 1980	12.00	October 05,1987	11.00
February 28, 1980	13.00	November 09, 1987	10.50
April 03, 1980	14.00	February 01, 1988	9.50
April 28, 1980	13.00	April 04, 1988	10.00
May 15, 1980	11.50	May 23, 1988	10.50
August 20, 1980	12.00	June 02, 1989	10.00
September 22, 1980	13.00	July 17, 1989	9.50
November 24, 1980	13.50	February 23, 1990	10.00

Source: First South Bank - Mortgage Lending Office, Orange Park, FL

Highlights of Chapter 10

• During the design process, your wish list can become very long and drive you over budget. This is because small priced options do not seem expensive by themselves, but become very expensive when added together.

• When you do go over budget, look at each optional extra for your home and find a reasonable compromise between what you want and what your budget can handle.

• As you did when first designing your home, concentrate on your home's layout, traffic patterns, and room sizes and find areas where you can scale back your design without sacrificing the functionality of your home.

• Consider cutting back on the finishing touches in your home initially and adding what you truly want later on when money is not as tight.

• When your builder determines your home's cost, the price should include a grand total of all costs, both large and small.

• Do not make the classic mistake of thinking you will be able to cut costs during your home's construction because **IT WILL NOT HAPPEN**.

• Remember that it is not unusual to have to redesign your home as many as five different times until you get the home you want and can afford.

• Waiting to buy a home will make affording your home harder, not easier. This is because the affordability gap, the gap between average income and new home prices, becomes wider with each passing day.

• When you build a new home today, you lock in your sales price today and pay for your home with tomorrow's dollars. Then the rising prices of new homes starts working for you instead of against you as your home's value increases.

[i] "New Home Prices" http://nahb.com/facts

Notes

Chapter 11:
Contract & Finalization Of Design Process

By this point in the design process, we have accomplished quite a bit:

- We have set a budget for our new home.

- We have chosen an area and a homesite to build on.

- We have checked into methods of financing with mortgage loan officers.

- We have investigated numerous builders and picked the right one

- We have reviewed dozens of stock plans and picked the best one

- We have customized our stock plans to fit our individual needs and tastes.

- We have checked prices

- We have added and subtracted optional items to come up with the optimum balance of desirability and affordability.

- We now have a final price, design, builder, area, homesite, financing, and we are ready to begin building our dream home.

Now the hardest work is done and the rest of the process should be fairly easy. However, at this point I will interject a word of caution. Of all the mistakes I have seen in this business of construction, (and in more than thirty years I have seen enough mistakes to fill a book ten times this size), most have occurred or at least started at this point in the process. The reason for this is simple- this is the point where you, the homebuyer, may start to get complacent and wander away, thinking that your work is over.

This kind of thinking is perfectly natural because by this time everyone is just a little bit tired. After all, your Realtor® or site agent has worked hard helping you and now that you have the home you want, she is ready to relax. Your builder has covered all the bases with you and given you a complete price on your new home with all the optional upgrades you wanted. And you are positive that you have conveyed everything you want and expect in your new home to your site agent and builder. Now all that is left to do is complete the contract, finalize your home financing, and begin construction.

As we said in the very first chapter, many people think they will check the job as it goes along and make sure that everything goes right. Any details that have not been addressed can be taken care of during the building process as they come to light. Then they come back when the home is well underway and discover, to their horror, that many of the things they assumed were included or covered in their new home were not. Then they try to add these items late in the construction process and find that these additions are very costly and/or impossible to include at this late stage.

The construction of any new building, especially a new home, is very detail-oriented. Items that are unresolved and left to chance may become the source of the horror stories you hear so often. Therefore, the **most important step** in the entire design process **is to _finalize_ every facet of your new home design _NOW_, before you start building**.

Finalizing the design process is accomplished by using a good contract and having a final plan approval. It is imperative at this point that **EVERYTHING** pertaining to your new home must be **IN WRITING** and **SIGNED BY ALL PARTIES TO THE CONTRACT**. Whatever somebody told you was included or what you understood would be done will not count for **anything** if it is not clearly shown in the written contract or the approved set of plans. The adage about oral agreements being "not worth the paper they are printed on" never holds truer than when you are talking about a new custom home.

When you look at builders' contracts, you will find that there are probably as many forms and types of contracts as there are builders. These contracts cover a very complex section of the law known as real estate law. Most concepts in this branch of law are beyond the scope of this book (and to be perfectly honest, beyond the expertise of its authors). Therefore, the contracts we use in our business on a day-to-day basis have been reviewed many times by qualified real estate attorneys to ensure that they are legally correct, fair, and easy to understand.

Whatever type of contract you and your builder use, it should spell out any areas of concern you have or any special

> **Author's Note:**
>
> Only a real estate attorney licensed to practice in your state is qualifies to give you legal advice and evaluate the validity of the contractual agreement between you and your builder. The brief explanation of contracts we provide in this chapter is **not** intended in **any** way to **substitute** for the advice or review of a qualified real estate attorney. Our explanation is intended only to acquaint you with a basic understanding of what you can expect when you enter into a contractual agreement with your builder.

circumstances that apply to your home. Your contract is a "legally binding agreement", binding everyone who signs it to the terms and agreements which are stated in it. So until an item is written as part of your contract, it is not part of that agreement and your builder is not legally bound to abide by it.

Any time you are asked to sign a contract and you have any doubts or questions, you should take a copy of your contract to an attorney for their review and comments. Even though your attorney's fee may seem high, it in fact may be a bargain compared to the money, time, aggravation, and grief that could occur because of a mistake or a misunderstanding. No legitimate builder or real estate agent will object to you taking a copy of their contract to your attorney for their review. If they do, then you should **seriously reconsider** whether you have chosen the correct building professionals for your new dream home.

To acquaint you with some of the basic elements of a standard new home purchase and sale agreement, or contract, we have included a list and explanations of some of the elements we consider essential in our own contracts and which we expect for you to see in your own. This list is not intended in any way to substitute for the advice or review of a qualified real estate attorney; it is only intended to show basic contract elements such as:

- Names of the builder and owners as they will be shown on the final deed to the new home.

- Legal description of your new homesite property.

- Description of the work that is to be done (your new home).

- Total price of that work.

- Method of payment for the work (cash, mortgage).

- Closing costs, financing cost on permanent and construction loan financing.

- Start and completion dates for construction.

- Means of handling changes that may become necessary during construction, and

- Sole agreement clause which states, "This contract and any addenda is the sole agreement between the parties and supercedes all other agreements, oral or written, stated or implied."

Names

The names of the builder and homeowners may seem like a simple thing, but it is amazing how much trouble it causes if they are wrong or misspelled. A typical home here in Florida generates enough paperwork to fill the bed of a pickup truck. Papers from the mortgage company, title company, permits filed at the courthouse, and everything else connected with your home contains information that will come from your contract with your builder. So make sure your name is spelled out on your contract the way you want it to appear on your paperwork and the deed to your property. Because correcting a misspelling on all of these reams of paper could prove to be a daunting task.

Legal Description

A correct legal description of your homesite is important to ensure that your home is **NOT** built on the wrong parcel of land. Also, the title and mortgage companies need your legal description to complete their paperwork properly and make sure you end up with the title to your home instead of your neighbors'.

If you are building your home in a subdivision, the legal description of your property is easy to check. Just compare the lot and block numbers for your homesite with those on a recorded plat of your subdivision (you can find this at your county clerk of the court office) to see if they match. For example, if your homesite is Lot 100 Block 2 Dreamland Estates, you would get a recorded plat of Dreamland Estates Block 2 and look for lot #100. If this lot is the same as your lot, you are in business. If not, you and your builder will have to get the legal description corrected before signing your contract.

If, on the other hand, you have decided to build your home outside of a subdivision, then the legal description for your lot may be half of a page long and written in gibberish only a surveyor can understand. In this case, the legal description should be attached to the back of a survey which shows a diagram of your property. If this diagram does not look like your property, then you will also need to get this description corrected or possibly have your property resurveyed before you begin building your new home.

Several years ago I heard about a competing builder who was building near us in a subdivision and did not carefully check the legal description on the contract before laying out his clients' home. After having poured the concrete floor slab of the home he discovered that his superintendent had mistakenly centered the home directly on the property line between his lot and his neighbors'. So with thousands of dollars already invested in the building of the home, the builder found that half the home was on his lot and the other half was on somebody else's. Faced with a no-win situation, the builder did the only thing he could. He bought his neighbor's lot for substantially more than market value and then generously gave his homeowner two premium lots for the price of one.

Description of Work

In order to avoid any potential misunderstandings, it is extremely important that a complete description of the work your builder is doing be included with your contract. If your dream home is a builder's stock plan which you have modified, then it is easy to state, "Builder model plan # or model name as built on

Lot #100 Block 2 Dreamland Estates with the following modifications." This way you tie down the basic plan you are using and then list **ALL** changes to that basic plan you have made such as: room sizes, upgrades, elevation changes, and any other changes making this home uniquely your own. This list of changes can get very long; therefore, we usually attach an addendum or extra sheet to the contract, listing the changes and state "see addendum for changes to standard model." This way, there is no confusion about what is going to be built once construction starts.

Proposal/Description of Work

PROPOSAL FOR: Mr. & Mrs. John J. Client
Lot 100 Dreamland Estates
Clay County, Florida

STANDARD MODEL: Brittany 4A as built on Lot 99 Dreamland Estates

STANDARD PRICE: $320,615

MODIFIED AS FOLLOWS:

Full Brick Veneer all around	$20,000
Footage Stretch #1: Add 1'-0" to bedroom and garage wing = (37x1 ft) x $48/SF	$1,775
Footage Stretch #3: Add 1'-0" to garage (21'x1') x $20/SF =	$420
Footage Stretch #4: Add 3'-0" to rear of bedroom wing (27'x3')x$48/SF=	$3,888

MODIFIED MODEL HOME PRICE **$346,698**

Acceptance of this proposal by owner and receipt of binder by contractor shall constitute authorization for contractor to prepare finish plans, specifications, and contract. Binder deposit of: Ten – thousand & no/100 dollars ($10,000)

shall be deposited in contractor's escrow account and will be credited at closing against total costs as agreed upon including any extras that may by mutual agreement be added. If for any reason contractor cannot be given authorization to proceed within 30 days of execution of finish contract, cost of finish plan preparation shall be deducted from binder.

_______________________________   _______________________________
John J. Client   **Contractor's Representative**

____
Kelly A. Client

When your home is a completely custom home that your builder has never built before, it may be necessary to attach a complete set of blueprints as well as specifications to your contract. Although your copy of the contract will seem like an armful to carry home and store, it will help avoid confusion and be a ready reference tool for you should questions arise during the construction process.

Total Price of Work

This may seem obvious, but it is important that the total price you will pay your builder be listed on your contract. The base price of the model you are modifying should be listed, along with each modification and upgrade and the corresponding price of each. At the bottom of this list should be the total price of the base home and the sum of all the upgrades you want your builder to include. This total price will be the price you will pay your builder for your new home.

Some of you may be saying, "This list of upgrades could get pretty long. Wouldn't it be easier to just put a lump sum for all the extras in the contract and save space." Well, yes and no. While your sales agent is writing the contract, it would be easier to use a lump sum because it would initially save both space and time. However, as soon as a misunderstanding arose about what was included, what was optional, or the price of an individual option, then the time savings would disappear. Without a list attached to the contract, we would have to go back through the construction file and/or try to remember what we had agreed on. And since your memory may be different from mine, this kind of misunderstanding could become quite a source of conflict.

But when we have a list of each option and its corresponding price, there is no source of conflict. Although compiling this list requires some time, it provides an easy reference source should a question about extras arise. There is never a question about whether this or that item was part of the total price, because it is all there in black and white for you and your builder to see.

Method of Payment

As you can tell from the price of homes these days, construction projects large and small are very expensive to build.

Usually a builder is not able to carry a large number of construction jobs from beginning to end without receiving some kind of partial payment before the job is completed. These partial payments are obtained either from the buyer or the construction lender in the form of draws or progress payments. It is important to decide which method will be used to pay your builder so there is no doubt how payments are to be made and the amount each payment will be.

If your home and homesite are to remain in the name of your builder until the closing (as is the norm in a subdivision), then construction loan financing is usually handled by the builder. The construction loan lender makes partial payments directly to the builder without the buyer's involvement. If you are going to pay for the home in cash or build it on your own offsite lot, then you will serve as the construction lender to your builder, and you will include a draw schedule in your contract which will specify at what stages each partial payment will be due and what the amount of each payment will be.

RESIDENCE FOR MR. & MRS. JOHN J. CLIENT
DRAW SCHEDULE

CONTRACT PRICE	**$346,698.00**

DRAW #1:
Floor slab poured & finished; foundation survey
20% or $69,340 & any extras added to date — **$ 69,340.00**

DRAW #2:
Framing complete including roof & wall sheathing; roof dried in;
Plumbing topout complete & tubs set; electrical rough complete
Air Conditioning ducts in place
40% or $138,680 & any extras added to date — **$138,680.00**

DRAW #3:
Drywall complete; trim complete; cabinets set; roofing complete;
ceramic tile in baths set
20% or $69,340 & any extras added to date — **$ 69,340.00**

DRAW #4:
Substantial Completion – Exterior & interior ready for final
inspection; landscaping, exterior concrete, finish flooring, all
mechanicals set and ready for final inspection
15% or $52,005 & any extras added to date — **$ 52,005.00**

DRAW #5:
All inspections complete; walk-thru and final punchout items
complete; all mechanical systems checked out and operational;
all cleanup complete; ready for move-in; final survey; waiver of
lien affidavits from any subs or suppliers sending notice to
owner; builder's final affidavit; warranty documents
5% or $17,333 & any extras added to date — **$17,333.00**

Closing and Financing Costs

It is very important (and what in this section isn't) that the issues of closing costs and construction loan financing costs be addressed on your contract. In most contracts, new homes are built and transferred to their owner in one of two ways. The first method is the one which most people think of where the homeowner acquires a piece of land and retains the title to that land while the home is being built. In this traditional method the builder typically provides only construction services. All of the financing costs, including the cost of construction loan[i] financing, are paid for by the owner and are not included in the builder's fee. However, when you negotiate with your builder, you can either ask him to include these financing costs in his contract or you can pay for them out of your own pocket. Either way, a construction loan will almost always be taken out on the property because construction costs must be paid while the work is being done and a permanent mortgage does not take effect until the home is complete.

Construction loan financing on a new home is usually provided for six months to a year and the funds are released as the construction of the home progresses. As with all loans, interest must be paid on the money borrowed, so your contract should state whether you or your builder will pay this construction loan interest. Also, you must decide which of you will pay the costs of converting this construction loan to a permanent mortgage when your home is completed.

As soon as your new home is complete and the builder is ready to turn it over to you, the closing will be held and your permanent loan financing or mortgage will be placed on your home. At the closing, the closing agent, usually a title company or an attorney, will collect the construction loan interest costs and the costs associated with the closing (from you or your builder according to the terms of your contract). The closing agent will use this money to pay off the construction loan, the closing costs, and any additional money due to the builder. Then the permanent loan on your home will be recorded in the official records of your local courthouse. Your property and home then serve as collateral on the mortgage, and you will start paying normal monthly mortgage payments to maintain the loan in good standing. As my lawyer explains to my clients, the permanent loan paperwork you sign at the closing boils down to one simple phrase, "If you do not pay, you cannot stay."

The second method of building a new home and transferring it to the owner is the builder retains the title to the property throughout the construction process. When the home is completed, the permanent loan is put into place at the closing and the closing agent pays off the construction loan with money from the permanent loan. The builder then transfers title to the owner and the owner begins making the monthly loan payments, knowing that "If he does not pay, he cannot stay."

This second method is more common for new construction in a subdivision. The builder arranges for the construction loan on the property and pays all the construction loan costs without involving the homeowner. The builder in a subdivision also pays some of the permanent loan or mortgage costs; however, since each builder will pay different closing costs, the contract should specify which of these the builder will pay and which the homeowner will pay. When you buy in a subdivision, your site agent should prepare a closing cost estimate which illustrates the closing costs the buyer and builder will pay at closing. It is important that these costs are noted in your contract so you will know how much money you have to bring to the closing table.

Start and Completion Dates

The construction of your new home depends on a multitude of factors including weather, materials, supplies, and personnel. These factors can and often do combine to cause delays in the completion of your home. Such delays make it impossible to completely tie down a construction schedule because construction events are constantly being adjusted. Thus a construction schedule can be thought of as more of a suggestion than something chiseled in stone. When you and your builder set start and completion dates, remember that these dates are approximate and expect them to vary anywhere from a couple of weeks to a month or more.

Construction Changes

The idea of making changes during construction is one of the least understood concepts in the entire industry. Many people mistakenly believe that when they want to make a change to their home, all they have to do is call into the main office or run by the jobsite and tell whoever happens to be there to change it. Then their changes will be done just as they requested with no questions asked.

About ten years ago one of my homebuyers wanted rustic cedar lap siding on his home, and on his approved plans he requested that it be run straight across his walls, parallel with the ground. Soon after we had installed the siding, the buyer called stating that he wanted the siding over the garage door run at an angle, parallel with the roof slope. I told him that his approved plans called for it to run straight, and it was done according to these plans which he and his wife had signed, and thereby approved.

"Well, I called your office the other day and told you guys I wanted it changed," he replied.

"Who did you speak to?" I asked.

"Well, somebody."

"Who was it specifically you spoke to?" I inquired.

"I don't know. Whoever picked up the phone."

He did not know to whom he had spoken and neither did anyone else in our office. Maybe he spoke to the janitor? If he did, I hope that he did not expect the janitor to have the authority to make changes to his home. Because if anyone, including your builder, can so easily change your home after you have specified exactly what you want built, I guarantee you will not be happy when you see the finished product.

Because a policy of making arbitrary changes at will to your home is not feasible, your contract will probably have a clause specifying how to make changes to your home once construction has begun. Every change to your home should be in the form of a change order, and each should have the following characteristics:

– The order should be in writing and signed by all parties to the contract (owner and builder or his authorized representative)

– The order should describe the work in detail so that the owner and builder know exactly what is to be done

– The order should state the cost of the work along with an explanation of whether that cost will be added to or subtracted from the sales price

CHANGE ORDER

RESIDENCE FOR: **Mr. & Mrs. John J. Client**
Lot 100 Dreamland Estates

DATE: **Friday, July 23, 20xx**

The following changes are hereby authorized to be added to the above captioned job in accordance with the original contract by and between XYZ Builders, Inc., as contractor, and John & Kelly Client, as owner. Additional costs payable in accordance with original contract terms.

1) Add extra telephone jack in master bedroom.

    ADD: **$100.00**

2) Add second light switch to dining room fixture for
    future ceiling fan.

    ADD: **$100.00**

3) Do **NOT** provide light fixtures for dining, master bedroom,
    or the two spare bedrooms.   Fixtures will be provided and
    installed by owner.

    ADD: **No Charge**

TOTAL ADDITIONAL COST: **$200.00**

_______________________________   _______________________________

Mr. John J. Client   Contractor's Representative

____

Ms. Kelly A. Client

If the change order is not in writing and signed by both the builder and the owner, then it may not be considered part of the contract, and there is no guarantee that the changes ordered will be done. Most people believe that this is just a safety measure to protect the builder, and they are absolutely right. A builder must have this protection, or else it would be impossible to make changes of any kind. For example, if the owner came by the jobsite and told the superintendent he wanted something changed, then decided a week later he did not like what was done, then the builder would be forced to convert the change back to the way the approved plans originally showed it. Without a change order, the builder would have no documentation to prove the owner requested the change, and he would be legally required to provide what was documented in the contract and the approved set of plans. This means that without a written change order, the builder would be forced to change the home back to what was on the approved plans **at his (the builder's) expense**. Thus, a change order is necessary to legally protect the builder.

But while a change order protects the builder, most people do not realize that **a change order protects the homeowner even more**. For instance, without a written change order, there is absolutely nothing to stop your builder from extorting money from you at the closing table. Just as the papers are about to be signed, he can lean over to you and say, "By the way, we can close on this home as soon as you pay the $5,500 in extras you authorized."

"But we authorized no extras!" you might exclaim.

"Yes, you did. You called my secretary last month and asked for these changes to be done. So until you pay for them, we are not closing on this home."

In the meantime, you and your family have been living in a hotel for a week and a moving van full of your furniture sits on your driveway. Without some kind of documentation to support your case, you may be forced to pay your builder thousands of dollars for changes that you never wanted. By using detailed, written change orders, you and your builder can avoid such potentially harmful situations and protect yourselves and each other.

There is another important reason why construction changes

must be done only in strict accordance with the contract and properly documented by a change order. Any change made by someone not authorized to make that change or made without proper documentation may **VOID YOUR NEW HOME WARRANTY** on that portion of your home. For example, several years ago one of our owners came into her home at the very end of the job and, without saying a word or documenting a change, she added a microwave range hood over her kitchen stove. Now adding a combination range hood and microwave over your stove is perfectly fine– in fact, we install them frequently for our clients. However, when you add one, it is usually necessary to add a dedicated electrical circuit exclusively for this microwave range hood. This is because the microwave hood uses so much power it can cause the circuit breaker for your kitchen to overload and shut off or "trip" every time you use it. Unaware of this, the owners hooked their microwave into the kitchen circuit with no authorization and patted themselves on the back for saving the $75 installation charge. But after moving into their home, they found that every time they used their microwave and another appliance in the kitchen simultaneously, they tripped the circuit breaker. The electrician inspected the problem and told them, "This wiring [for the microwave] was not done by us. Therefore, we will not warranty this work and cannot fix this problem."

The irate homebuyer then called us, but because they had installed the microwave without any authorization or documentation, they had, in fact, voided their warranty, and there was nothing we could do. Soon after this call, the homeowner reported the electrical contractor, and us, to the local building department and made a formal complaint. The local building department investigated and concluded that "the owner acted illegally in making the change without proper documentation and without using properly licensed personnel to do work." Therefore, neither we nor the electrical contractor were legally responsible for correcting the homeowner's problem.

These homebuyers thought they were saving money, but after getting another electrical contractor to run a dedicated circuit for the microwave through their attic, they probably ended up spending several hundred dollars. When you compare this with the $75 fee we would originally have charged them, the benefits of getting proper authorization before making a change to your home become obvious.

Another similar situation happened several years ago when an owner made several modifications to his electrical system <u>after</u> he moved into his home. When problems occurred with the electrical system, the electrician inspected the home, discovered the wiring changes, and refused to do any further work to the home because by tampering with the electrical system, the owner had voided his electrical system warranty. The owner was then forced to hire someone to help him correct his modifications to make his system work correctly again.

As it turned out, this particular owner held a degree in electrical engineering and, therefore, should have logically been qualified to make the electrical changes himself. But having an electrical engineering degree is not the same as being a licensed journeyman electrician. Changes in electrical work as well as other mechanical work have to be done by professionals who are licensed contractors. In most cases, an electrical engineer is not a journeyman electrician; therefore, he is not allowed <u>by law</u> to change the wiring in his home.

As in our first example, if this homeowner had followed proper procedure and made the changes he wanted in accordance with his contract, he could have avoided a lot of trouble and gotten the electrical system he wanted. Only by following proper procedures and making changes to your home properly can you be sure that the home you want is the trouble-free home you will receive when construction is complete.

Sole Agreement Clause

You will probably see a sole agreement clause at the end of your contract which states that the contract, the approved plans, the specifications, and any addenda to the contract constitute the sole and complete agreement between the owner and the contractor. Although this sounds simple, I cannot tell you how many times homeowners have wanted to add to the plans and specs after the building has begun. Clients have often come to me and said, "Well, we talked about including this," or "I clearly remember us talking about doing that," or "I thought we mentioned using this, not that." The plain truth is when you go through the long, involved process of designing a custom home and working out all of the numerous details, you will discuss countless options and possibilities. There is absolutely no way you or your builder could remember all of them. So it is important that everyone relies on the contract and all of its addenda to serve as the sole and complete agreement between the builder and the owner. Of course, legitimate change orders that follow the procedure outlined in the contract will become part of the contract. But outside of these properly done change orders, nothing else can alter your written contractual agreement in any way.

Once the contract is completed and signed by all parties, the home plans must be finalized through a plan approval session. During this two or three hour period, the builder, the owners, the site agent, and the Realtor® (if involved) will discuss every aspect of what is to be built in the home. From electrical switch placements to the materials used throughout your home, everything pertaining to your new home will be thoroughly reviewed so that by the end of the plan approval you will have no doubts about what is included in your new home.

We usually start off the plan approval by reviewing the signed contract and making sure that everything in the contract was included in the approved set of plans. Then we run down the owners' list of concerns and questions and make sure that any items that have come up since the signing of the contract are fully addressed. Finally, we run down our checklist and discuss every item, whether it was changed or not, as a final check on the plan and make sure that there are no misunderstandings as to what is being built.

As you can tell from our plan approval checklist in the back of this chapter, our list has evolved over the years to cover almost every item that must be reviewed during the plan approval process. Although we constantly update this list, we know it will never be complete simply because every home is different, and no list could cover every facet of every home we will ever build. Our list does, however, cover most of the basic items that need to be discussed with every home. While going over this list, it is important that the sales agent and the designer be present so they can check their notes against the contract and plans and make sure that nothing was accidentally omitted.

Any changes made to the plans during the approval process are marked in red ink so we can easily modify the final home plans to conform to the approved set. On occasion, if there is an error in the plans or some major change is requested, you may have to reschedule the plan approval so that your builder can redraw the plans. Some people are hesitant to do this because redrawing their home plans will lengthen the already tedious plan review process. If this happens to you, remember that taking more time for your plan approval is perfectly acceptable. However, cutting your plan approval short **is not**.

You must **<u>never shorten the plan approval</u>** process for one simple reason- this is the **<u>one</u>** and **<u>only</u>** way to insure that the home you are going to receive is indeed the home you want. It is no coincidence that the people who are happiest with their homes are the ones who took the longest time at their plan approvals and asked the most questions. This is the point at which everything, and I do mean **<u>everything</u>**, is finalized.   The reason why we go to so much trouble finalizing everything is because anything not finalized is guaranteed to pop up and create a problem somewhere down the line.

The reason why finalization is such a critical item is simple. During the plan approval the homebuyers sign the house plan, indicating that they approve of what they see and this is exactly what they are paying the builder to build.   The designer then redraws the final blueprints from these approved plans to reflect any changes made during plan approval. These final blueprints are then copied and used to obtain permits from local authorities and are given to every person who will work on your home. If any changes are made to the approved plans after they have been handed out, a new, revised set of plans must be given to each of these people, and the old plans must be destroyed to avoid confusion and mistakes.

With the vast number of people working on your home and the great number of plans floating around, it is difficult to make sure that the changes you have ordered get to the right people. Any changes you make after the plan approval process greatly increase the chance for mistakes to be made. The process of notifying all the people involved with your home's construction about the changes you have made is very costly and time consuming. This is a major reason why changes to your home during the construction process are expensive and why you must finalize everything during your plan approval.

My standard line at the end of every plan approval session has always been, "What you see here is what you get. If what you see on these plans and in your contract is what you want built, then you will be very happy with your new home. However, if what you see here is not what you want built, then you will not be happy with your new home." At every opportunity, I stress the fact that what you see on your approved plans is what will be built, so it is imperative that you make sure what is printed on those plans is **exactly** what you want. If it is not, you are guaranteed to have trouble farther down the line.

I remember several years ago where we finished a plan approval session in about two hours and after I gave the homeowners my standard "What you see is what you get" spiel, they laughed, signed the plans and walked out the door. As I sat talking with the site agent, I heard their car start, idle a few moments, and then shut off as they rushed back through the front door with their smiles completely gone. Looking at me the husband explained, "While we were sitting in the car, we suddenly realized what you had said. We think we had better have another look at those plans just to make absolutely sure that is the home we want."

"Absolutely," I responded jubilantly. "Come on back. Spend all the time you need and make sure this is exactly what you want."

They made no changes to their approved plans that night, but they left my office knowing that they were going to get exactly what they wanted. Their job went smoothly after that, and they were one of the best customers I have ever had. If you want to make sure your home will be right, just remember: what you see is what you get, so make sure that what you see is what you want.

I am always amazed at how lightly the plan approval process is taken by many owners and even some salespeople. As a builder, I consider the plan approval to be the single most important part of the design/build process for one simple reason– the plans are the instructions to the people who are going to work on your new home.   Neither the builder, the owner, nor the salespeople are going to be on the job the majority of the time your home is being built. Any work done on the job will be done in accordance with what is printed on the plans. Therefore, if the plans are correct and show what you intend to have built, everything will be fine. However, if the plans are not correct, then the home will be an absolute disaster. What is shown on the plans, **not** what is in your mind, will be what is built in the field.

When you are building a completely custom home rather than modifying a builder's standard model, your plan approval will take place prior to the execution of the contract. This is because your custom plans and specifications must be attached to the contract in order for the contract to properly and completely describe the work to be done.

When you modify a builder's standard model, the plan approval can take place after the contract is signed. In this case, there should be a clause in the contract stating, "Contract is subject to plan approval" so that any misunderstandings can be corrected by all parties during the plan approval.   If these misunderstandings are not corrected, then the contract will be null and void. This is your builder's "guarantee" that any mistake up to that point will be corrected before your plans are finalized or your home will not be built.

After the plans are approved and the contracts are signed and legal, it is time for the last step in the finalization process– the color selection. Although this is the last step, it is by no means the least. Your interior and exterior color selection choices can alter your home design because you may not only determine the colors of your home, but also change some of the materials used in your home as well. For example, you may decide that a stucco home is not nearly as elegant as the brick home you saw in Atlanta a couple of years ago and change your exterior from stucco to brick. If you decide to make such drastic material changes, you should do so as soon as you can because most home materials have to be ordered early in the construction process to avoid unnecessary delays.

And that should do it. Once your contract is complete and signed, the plan approval is done, your approved plans are signed, your color selections are finalized, and your color selection sheet is signed, the design process is officially over. Now all you have to do is let the builder get to work and watch your ideas take shape as the next several months transform your vacant homesite into the home of your dreams.

Plan Approval Checklist

Owner :   John J. Client
Model: Brittany 4A modified
Job Number: Lot 100
Subdivision: Dreamland Estates

Site Plan

Size of homesite
Orientation of home
Layout of home on homesite
Easements on property (if any)
Type of drainage (A,B,C)

Exterior Concrete Paving

Layout of driveway
Location of patio, front & rear porches
Location of sidewalks
Air conditioning compressor pad location
Any extra concrete owner requests (if any)
Special paving considerations unique to this homesite (if any)

Exterior Home Walls (material, size, appearance, door & window openings, etc.)

Front exterior wall
Rear exterior wall
Side walls of home
Special Trim Details (porch railings, decorative woodworking, etc.)
Porches
Balconies
Exterior doors & front entryway details

Roofs

Roofing material (shingles, metal, tile, slate, etc.)
Roof pitch or slope
Location of roof vents
Special roof flashing or trim (if any)
Fascia and soffit material

Plan Approval Checklist (continued)

Windows

Size of each window
Type windows (colonial, clear glass, etc.)
Window glass glazing (obscure, clear, tinted)
Window frame material (aluminum, wood, plastic, etc.)
Window frame color
Special window details
Window sills (material, appearance)

Room Finishes

Interior walls (painted, paneling, etc.)
Ceiling types (vaulted, level ceilings)
Ceiling heights
Special trim details (art niches, special moldings, etc.)
Interior trim packages (type trim moldings)
Interior trim finish (stain, paint)
Fireplace type (gas, wood-burning)
Fireplace elevation
Plant shelves (if any)
Door type (hinged, bifold, pocket, etc.)
Door finish (stain, paint)
Interior door style (Colonial, Luan, Classique, etc.)
Style of door locksets
Deadbolt locksets
Special door hardware (if any)
Closet shelving
Linen closet shelving
Kitchen pantry shelving
Tub enclosures (if any)
Plumbing fixtures (bathtubs, showers, toilets, etc.)
Stairs & Balconies (treads, risers, railings)

Cabinets

Kitchen cabinet layout
Kitchen counter tops (material)
Kitchen counter backsplash (material)
Layout of bath cabinets
Knee space in bath cabinets (if any)
Bath mirrors
Medicine cabinets
Bookcases / Built-in furniture / Entertainment Centers

Plan Approval Checklist (continued)

Appliances

Range & oven or cooktop stove
Range hood
Built-in microwave range hood (if any)
Garbage Disposal
Kitchen sink
Bathroom sinks
Dishwasher
Special equipment

Finish Flooring (type, location)

Carpeting
Vinyl Flooring
Wood Flooring
Other
Flooring Breaks

Ceramic Tile

Floor tile size and location
Bath tile size
Bath tile around tubs and showers
Kitchen tile – floor tile and counter tile and backsplash (if any)
Foyer tile – size and style (if any)
Other (if any)

Electrical

Phone outlets (number and locations)
Light switch locations
Cable TV outlets (number and locations)
Receptacle cover colors and style
Ceiling fans (number and locations)
Interior lighting (type and locations)
Exterior lighting (type and locations)
Specialty lighting (recessed, floods, etc.)
Electrical service amperage
Exterior receptacles
Specialty items (if any)

Highlights of Chapter 11

- This is the point in the construction process where most mistakes start because you, the homebuyer, think your work is done.

- All that is left to do at this point is complete the contract, finalize your home financing, and begin construction.

- Many people think they will check the job as it goes along and correct any unresolved items as the home is being built. Then they find that features they assumed were included were not, and now that the home is well underway, these features will either be very costly to add or cannot be added at all.

- The **most important step** in the entire design process is **to finalize every facet of your new home design NOW,** before you start building.

- **EVERYTHING** pertaining to your new home must be **IN WRITING** and **SIGNED BY ALL PARTIES TO THE CONTRACT.**

- There are probably as many forms and types of contracts as there are builders. Whatever type of contract you use, it should spell out any areas of concern you have or any circumstances that apply to your home.

- Until an item concerning your home is written as part of your contract, it is not considered part of the agreement between you and your builder and your builder is not legally obligated to abide by it.

- Any time you are asked to sign a contract and you have any doubts or questions, take a copy of your contract to an attorney for their review.

- No legitimate builder or real estate agent will object to you taking a copy of your contract to an attorney. If they do, you should **seriously reconsider** your choice of building professionals for your new dream home.

- Make sure your name is spelled on your contract the way you want it to appear on your paperwork and the deed to your property.

- Check the legal description of your homesite to ensure that your home is not being built on the wrong parcel of land.

- To avoid misunderstandings, be sure to have a <u>complete</u> description of the work your builder is doing be included with your contract.

- The total price you will pay your builder for your new home must be written in your contract. This price includes the base home price as well as any upgrades you are adding to your home.

- Partial payments to the builder during a home's construction are paid either by the homeowner directly or a construction lender. You must decide which method will be used for your home so there is no doubt how payments are to be made and the amount each payment shall be.

- Your contract must state which closing costs you will pay and which your builder will pay. Also, you need to clarify who will pay the financing costs, including the cost of construction loan financing, for your home.

- When you and your builder set start and completion dates for your home, remember that these dates are approximate and will vary anywhere from a couple of weeks to a month or more.

- Every change done to your home after the contract is signed should be done with a change order. A change order must be in writing and signed by both the builder and the owner or it may not be considered part of the contract.

- Change orders legally protect both the builder and the owner.

- Using a change order prevents unauthorized personnel from making changes to your home. Unauthorized changes can void your new home warranty on the portion of your home that has been changed.

- Everyone relies on the contract and all of its addenda to serve as the sole and complete agreement between the builder and owner.

- Outside of properly done change orders, nothing can change or alter your written contractual agreement in any way.

- After the contract is signed, the home plans must be finalized through a plan approval session where all aspects of your new home are reviewed thoroughly.

- The sales agent and designer must be present during the plan approval so they can check their notes against the contract and plans and make sure nothing was omitted.

- The plan approval process must never be shortened because this is the one and only opportunity to make sure the home which is built is the home you want.

- Everything must be finalized at the plan approval because changes made to your home during the construction process are costly, time consuming, and sometimes impossible to perform.

- The plan approval is the single most important part of the design/build process because what is shown on your plans, not what is in your mind, is what will be built on your homesite.

- Your color selections can alter your home design because you may change your choice of building materials as well as determine the colors for your home.

- Like all changes, if you decide to alter the building materials for your home you must do so early to avoid unnecessary construction delays.

[i] construction loan– short term mortgage put on property to fund construction costs. Funds are released to builder as the stages of construction are completed. Loan is paid off by permanent mortgage put on home at the closing when home is transferred from builder to homeowner.

Notes

Notes

Chapter 12:
Building the Building

Finally, we are at a point in the building process where we can see things beginning to happen. Dirt is being moved, concrete is being poured, the frame is leaping up from the concrete slab, and the piles of construction materials are beginning to look like a home. At this point everyone, including me, starts to get a little excited. I have to admit that ever since I was a little kid, this facet of the job has captured my imagination. It is very exciting to be a part of the design process and build the home on paper. Then, after all the fine tuning and adjustments are complete, I love being able to drive to the jobsite and watch my design evolve from a load of concrete and a pile of lumber into a home where someone's family will live and grow for years to come.

I think that most of our homebuyers share my enthusiasm for watching their home come together. However, the building process tends to be every bit as stressful for the buyers as the designing process because most people do not understand how a home is actually built. Every time they come out to the jobsite to check the progress of their home, they do not understand what they are seeing and become more nervous than they were before their inspection.

There is an age old adage that says the two things you never want to see made are laws and sausage. Whoever coined that phrase should have added building as a close third. The reason for this is simple; there is no such thing as a perfect job. Neither I, nor anyone else I have known during my years in this business, have ever seen a perfect job. If any builder tells you that their work is absolutely perfect, they are either deluding themselves or they have a very loose definition of the word "perfect."

Since there is no such thing as a perfect job, you should not expect one. Instead you should carefully inspect the model home of your perspective builder. This will give you a good indication of the quality you can expect in your home because your builder's model home is used to attract potential clients. Since no builder would display his second best to solicit new business, you can expect your home to be built as well as your builder's model, but no better. If you find the builder's model unacceptable, you should not choose that builder.

When the time comes to begin your home's construction, you must remember that a ship can have only one captain. When you picked your builder, you picked the captain who will command the construction process. It is his job to take your home from paper to finished product, **NOT YOURS**. This is why it is so important to make sure your builder is qualified to command your ship and pilot it safely to its destination. If you have chosen carefully, then your home's construction should be nothing less than smooth sailing.

Now that your research is over and your builder is ready to begin, the best thing you as a homeowner can do is **STAY AWAY!** My happiest clients have always been those who left town when I built their homes. Instead of inspecting their homes before they were completed and causing themselves unnecessary worry, they did exactly what everyone having a home built should do. They stayed away until the home was complete and **then** inspected their home during the walk-thru. In fact, some builders I know are so adamant about their buyers not seeing their home's construction that they have written a clause into their contracts banning their clients from the jobsite during working hours. Although I can understand where they are coming from, I have never put such a clause in my contracts because I believe it is legally and practically unenforceable and it creates fear and animosity in a homebuyer instead of alleviating it.

People who do not understand construction will often see something that is a small problem and blow it out of proportion. Almost always, the problems they find are either too small to be worth mentioning or will be fixed later down the road. Either way, had they stayed off the jobsite, they would never have noticed the problem and would not have worried needlessly.

Now, I have just spent a good deal of time telling you not to visit the jobsite until your home is complete and ready for the walk-thru. However, since almost none of my former clients have heeded this advice, I doubt that many of you will either. For most people, a home is the single biggest investment they will ever make. So it is only natural for you to want to make sure that you are getting what you are paying for. So do yourself a favor. To maintain your peace of mind, make sure you research your builder thoroughly before signing a contract (something we will say again and again because it can never be overemphasized) and limit your number of visits to the jobsite. Doing these two things will keep your blood pressure down and make your home building experience a much more pleasant one for you and your family.

One thing you must be wary of at this stage in the construction process is the well meaning but uninformed friend or friends who always come out of the woodwork with abysmal horror stories and words of warning. After visiting your home one day, you may become excited and tell a friend, "I was just looking at my house, and they have started framing and should have the roof on by the end of the week." Whereupon your friend, after a dramatic pause will caution you, "You better watch out! My friend had a house built and this happened and that happened and they never saw their builder again." And before you know it, your excitement has been replaced by paralyzing fear in thirty seconds flat.

During these times remember that you did a lot of research to find your builder, and like the captain of a ship, he is qualified to command this construction voyage right to the end. Sometimes you may have to take this on faith, but if you have been careful, followed our advice, and did your homework, you should trust the builder you chose and let him do the job you hired him to do - build the trouble-free dream home you have always wanted.

Before you begin construction, your builder will obtain the building permit for your home in his name. This means that he is the "contractor of record" with the local government and takes responsibility for everything concerning the building of your home. Since he is taking absolute responsibility for **every** part of your new home, he **must** have the **absolute authority to control the job** to ensure that your new home is done correctly.

Partway through the construction process, many people get scared and decide that they want to supervise the building themselves. With some especially troublesome clients, I would have loved to turn the job over to them just to get them out of my hair. There is only one problem– it is illegal. The contractor of record is legally responsible for your home, and he is legally bound to build your home until its completion. Therefore, since you cannot change builders in midstream, you must thoroughly research your builder so you can decide whether he is competent to build your new home **before** he builds your home.

Probably the single biggest aspect of new home construction that surprises many people is that the workers you see throughout the day do not work for the owner of the home. In fact, they often do not even work for the builder. Most of the time, the construction personnel you see work for a subcontractor who in turn is hired by the builder to do a construction activity. This subcontractor has the same basic kind of contract with the builder that the builder has with the owner. Therefore, when you go out to view your home's construction, the workers you see do not work for you and will not regard you as their boss. They will only listen to the builder or to their superintendent if they are subcontracted labor. So do not be offended if the workers you see do not follow your instructions. They have already been instructed by their bosses, and if these workers were to accept conflicting commands or suggestions from you, then your home could be built incorrectly.

Now as we said earlier, the owners who are on the job least are the happiest with their finished home. However, we all know that if you are in town while your home is being built, there is no power on earth that will keep you away from that house. So since you are going to be visiting your home on a regular basis, it is important to tell you about some of the things you will see as your new home progresses.

I have heard construction jobs described as "organized chaos" and many people would question the "organized" part. When you go there, you will see piles of trash, piles of material that may look like trash, and people running around doing a lot of different things you may not understand. And for some owners, seeing their dream looking like a disorganized shambles can be a very traumatic experience.

Years ago, we had a call from a Realtor® who represented a couple that had come from out of town to check the progress of their new home. The Realtor® was excited and high strung because the people were very upset with the way their new home looked. "I don't know why," I replied. "The job is just being framed up and everything is going according to schedule, so I do not know what the problem could be." The couple insisted on meeting with me, so we arranged a time and got together at their new home. As I got out of my car, I saw that the wife had been crying, and the husband's face was as red as a fire engine. "It looks like all the framing for my house was just thrown together," he started yelling.

"Well, at the framing stage nobody worries about aesthetics," I explained. "All the wood you see now is going to be covered over by finishing material, gypsum wall board and exterior brick in your case, so they put these homes together fast and furiously. They do not exercise the same kind of care you would find with trim work, fireplace mantles, or other highly visible woodwork because nobody will ever see it once the finish material is put over it. So the framing is connected together in such a way that it will support any structural loads that are placed upon it. And although it is not attractive, it will support your home and will be invisible under your finishing materials in a couple of weeks, so why should it matter how it looks?"

After a few moments the homeowners calmed down and returned six weeks later to find their home beautifully finished and ready to move into. Once they walked through and inspected their new home, they were as happy as they could be and they even referred several of their friends to us. In fact, one of these friends told me why my client had been so flustered when he saw his home during the framing stage. Apparently, he had experience in finish carpentry and trim work on multi-million-dollar custom yachts along the east coast. He was accustomed to working with teak and other special woods that were so expensive they had to be handled as carefully as a jeweler would handle precious stones. After that kind of experience, seeing his home framed with wood that was full of knots and scratches was quite a shock. But once he talked to me and realized that framing lumber does not need to be as pretty as trim work, he calmed down and everything turned out fine.

Although the construction phase of a new home may look chaotic to the average person, there is actually a logical sequence of events taking place. Each construction operation is done in its own sequence, and some activities have to be completed before others can take place. For example, electric wiring must be run through the walls before electrical outlets and light switches can be connected. Other activities, however, can be done whenever the builder wants to do them without affecting your completion date. For example, the builder has about a month after the concrete driveway has been poured and the exterior walls of the home are finished to complete your home's landscaping. He may landscape a model home early so that prospective buyers can see a beautiful yard as they drive by or he may wait during the rainy season for the ground to dry a little before he rakes the yard and puts in the sod.

When the builder and his superintendent make a job schedule, they take into account this sequence of events that must occur when your home is built. They plan everything ahead and order materials ahead of time, particularly items such as custom cabinets and prefabricated roof trusses which can require two to four weeks, or more, to be made and shipped to your home. Items are ordered well in advance so that the workmen have the materials they need when they arrive on the jobsite. So there is order in the chaos you see when your home is being constructed. You just may have to look closely to find it.

To acquaint you with some of the chaos you will see as you watch your new home being built, we have included a general construction schedule in the back of this chapter that outlines some of the activities you will see on the job and the order in which they will most likely occur. Although builders may do some of these activities in a different order, they will all pretty much follow the construction sequence we have outlined here.

Any experienced construction superintendent will tell you that the job schedule he composes and follows is only an educated guess as to when each operation will take place because there are so many variables that affect his work. Someone could have a flat tire and be late for work or heavy rains in the Georgia lumber forests could diminish the tree harvest and cause a shortage of lumber for your home. Weather, logistics, availability of labor and materials all play a part in making the job schedule seem like more of a good suggestion than a commandment chiseled in stone. The construction schedule is constantly having to be updated and modified to accommodate the rapidly changing situation on the jobsite.

Of all the reasons for construction delays, weather has to be the most common excuse your builder will give. Although weather can be a big scheduling problem, I think it gets more blame than it actually deserves. This is because everyone knows that you cannot pour concrete in the rain or do many other things when the weather is bad. So even when labor or materials are not available, your builder may tell you that your home is delayed by weather out of simple convenience.

Many times when we have had a subcontractor tied up on another job or some material was delayed, I have thought it would be nice to have a good, heavy rainstorm. That way when the owner came out and said, "Well, it looks like this rainstorm might hold us up a couple of days, won't it?" I could look him straight in the eye and genuinely reply, "At least a couple of days. But don't worry, we'll work as hard as we can and get your home back on track in no time." Of course, a few days' delay is easy to make up during the course of construction. Many reasons for delay can and do happen throughout the course of building your home. It is just easier on your builder to blame it on the rain than having to provide a long, detailed explanation on the multitude of other causes that inevitably affect the construction scheduling for your new home.

As I have repeatedly admitted, I know you are going to sneak out and see your new home even though I expressly told you not to do so. However, there is one hard and fast rule I will give you that must never be broken under <u>any</u> circumstances. For the love of God, whatever you do, **<u>STAY AWAY FROM THE JOBSITE DURING THE LAST TWO WEEKS OF CONSTRUCTION!!!</u>**

The reason for this is simple – this is the messiest and most chaotic stage of the building process. Your home looks like a college fraternity house after a homecoming weekend. It is an absolute mess. The carpet has just been unrolled and carpet scraps lay everywhere. Construction debris of every sort fill the bathtubs. Appliance boxes are strewn throughout the house. The walls are scratched and marked and screaming for paint touch up. And to top it all off, workers from every trade are trying to finish their jobs and do not care what kind of mess they leave behind for someone else to clear away.

This is the construction phase that I call "later," because during the construction period there are always small items which the workmen put off finishing until "later." When the final two weeks of the job arrive, "later" becomes "now." And with the home looking like this and workmen finishing this and installing that at a feverish pace, "now" is a really good time for the owner not to be "here."

Avoiding your homesite during the "later" stage is especially crucial when the summer months are upon us. The air conditioner has not been hooked up and the stifling 110 degree heat has an adverse affect on peoples' attitudes.

However, after the place is cleaned up and the scratches and scuffs are painted over and repaired, your home becomes absolutely beautiful. In fact, I remember one client who made a punch list, or list of items in his home that need correcting, during the "later" period that was 14 pages long. After we readied his house for the walk-thru and began our final inspection of the home, he ran all through his house feverishly exclaiming, "You guys did everything on my list! There is absolutely nothing wrong with my house!"

"That's what happens when we finish your home," I replied. "Haven't I been telling you everything would turn out okay all along?"

Another time we had an older couple who sold their home up North and retired here in Florida. They chose one of our models, did their plan approval, and returned up North to settle their affairs. Unfortunately, they came back a couple of weeks before closing to sneak a peek at their new home. When they appeared on the jobsite right in the middle of the "later" period, I thought they were going to have a heart attack. I have never seen as much fear transmitted between a Realtor®, site agent, and owners as I saw that day. I honestly thought I was going to have to call the paramedics, but fortunately everyone was able to calm down before we had a medical emergency.

A week later, the couple returned to find their home clean and in perfect shape for their walk-thru. It is truly amazing how much difference a final cleanup can make in your new home. Cleaning the bathrooms, vacuuming the carpet, wiping down the countertops, mopping the floors, all improve the look of the job beyond belief. When they saw their new home, they relaxed and said they wished they had listened to me and waited until the walk-thru to see their new home. "Well, I have never had a client yet who listened when I told them to stay away," I replied jokingly "So don't feel too badly about it."

Frequently when the owner visits the jobsite, especially with friends or relatives, he will look around and find something that bothers him and write it down. Then he will see something else and write that down. Next thing you know, the owner has a paper full of "defects" in his builder's material and/or workmanship that demands instant repair. So he and his family and friends jump in their car and speed down to the builder's model or main office and rush in to show their list of grievances to whomever is available.

This scenario has occurred many, many times to me and to every builder who has been in this business any length of time. These same events replay themselves over and over again because builders are usually building for people who do not understand the building process, see something they do not understand, and get scared. The truth is that in more than thirty years, I have received literally hundreds of lists of this kind. And of all of those lists, I have never yet had an owner find a single problem of any significance whatsoever. **<u>NEVER</u>.** Most of the time minor construction mistakes are caught by the superintendent or the subcontractor, or by the builder himself, as he checks the job during the appropriate times. Never have I seen a list with a problem that I did not know about already or that caused any problem whatsoever. At least **90% of the time** they contained things that were normal or expected occurrences and caused **no problem at all**. And frankly, every one of these lists wound up where all useless things wind up– in the trash can. Of course, I would never tell an owner this industry secret, but since I am not building for you, I can tell you what really happens. A builder who knows his business knows what is going on and stays on top of his work. So, if there is a problem with your home, he will know about it and will not need you to remind him of it.

Now, do not get me wrong. If you feel the need to tell your builder about something that concerns you, that is fine. Just remember that if it turns out that any corrections need to be done to your home, these items will be fixed. But they will have to be fixed in the correct job sequence. Therefore, some problems you find may not be fixed the instant you tell your builder about them.

I remember one job from several years ago where the owner found spots on the wall where the painters had stained the doors and trim molding and leaned them against the wall. The doors and trim stained the wall and the owner became so concerned about the multiple blemishes that he called in and told my secretary that they had to be corrected. When I heard this, I laughed and told her that of course the marks had to be corrected, but not until the trim was installed and the painter came back for his final touch-up of the entire home. The next day the owner called again and said those spots were still on the wall. My secretary explained to him they would be fixed on the final touch-up, but the next day he called again and she went through the same explanation again. Finally, on the fourth day, I took the call myself and explained to the owner that the painters would be back later to do all the touch-up work at one time. So rather than doing these four or five little places now, we should wait until the end of construction and fix every spot at once. Because to correct those spots now, the painters would have to:

- stop what they were doing,
- spend an hour cleaning their brushes and equipment,
- take another hour to drive to his home,
- set up all their paint equipment again,
- and for what? For five (5) minutes of touch-up work.

Now, the painters could paint these spots all at once or take dozens of these five minute touch-up trips every time the owner saw a different spot on the wall. But when the painters take dozens of trips, they expect to be paid for their time. And who wants to pay for two hours of a person's time and receive only five minutes of actual work? The owner agreed with me that the painters should do everything at one time, and when we were finished, he had a paint job he was proud to display to everyone who visited his new home.

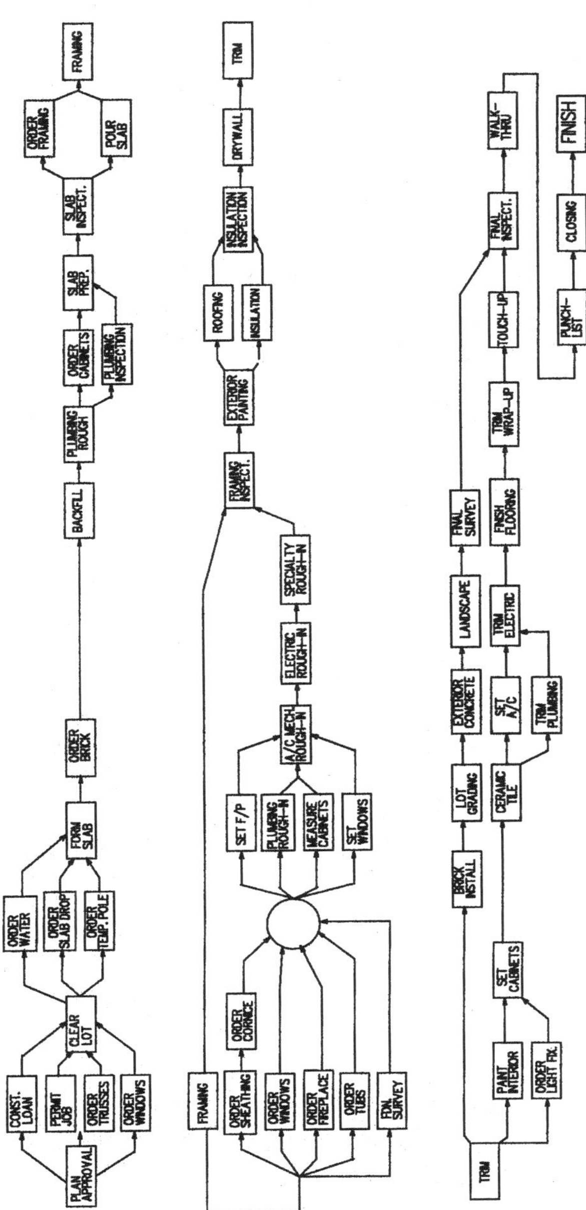
CONST. LOAN
PERMIT JOB
ORDER TRUSSES
ORDER WINDOWS
PLAN APPROVAL
CLEAR LOT
ORDER WATER
ORDER SLAB DROP
ORDER TEMP. POLE
FORM SLAB
ORDER BRICK
BACKFILL
PLUMBING ROUGH
ORDER CABINETS
PLUMBING INSPECTION
SLAB PREP.
SLAB INSPECT.
ORDER FRAMING
POUR SLAB
FRAMING
FRAMING
ORDER SHEATHING
ORDER WINDOWS
ORDER FIREPLACE
ORDER TUBS
FDN. SURVEY
ORDER CORNICE
SET F/P
PLUMBING ROUGH-IN
MEASURE CABINETS
SET WINDOWS
A/C MECH ROUGH-IN
ELECTRIC ROUGH-IN
SPECIALTY ROUGH-IN
FRAMING INSPECT.
EXTERIOR PAINTING
ROOFING
INSULATION
INSULATION INSPECTION
DRYWALL
TRIM
TRIM
PAINT INTERIOR
ORDER LIGHT FIX.
SET CABINETS
BRICK INSTALL
LOT GRADING
EXTERIOR CONCRETE
CERAMIC TILE
SET A/C
TRIM PLUMBING
LANDSCAPE
TRIM ELECTRIC
FINAL SURVEY
FINISH FLOORING
TRIM WRAP-UP
TOUCH-UP
FINAL INSPECT.
WALK-THRU
PUNCH-LIST
CLOSING
FINISH
ACTIVITY FLOW CHART FOR MONOLITHIC FOUNDATION HOME

Highlights of Chapter 12

- There is no such thing as a perfect construction job, so you should not expect your home to be perfect.

- Inspect the model home of your prospective builder and see if the quality of this home is acceptable. You can expect your home to be the same quality as your builder's model home, but no better.

- Remember that your builder will be in charge of building your home, not you. Therefore, you must research your builder thoroughly.

- When your builder is ready to begin construction, the best thing you can do is stay completely away from the jobsite while your home is being built.

- People who do not understand construction will often see a small problem and blow it out of proportion. Almost always, the problems are either too small to mention or will be fixed later down the road.

- To maintain your peace of mind, research your builder thoroughly and limit your number of visits to the jobsite.

- At this stage, be wary of well-meaning but uninformed friends who come to you with abysmal builder horror stories. Remember that no matter what they say, you can have faith your builder will build your home correctly because you thoroughly researched him before choosing him and know his work is first rate.

- Because your builder takes absolute responsibility for all facets of your home, he must be allowed absolute authority to oversee your home's construction. Therefore, since you cannot change builders or take charge of the construction process once your home is begun, you must research your builder thoroughly before building.

- The workers on the job will not accept commands from you because they have already been instructed by their bosses on what to do and if they accepted conflicting commands from you, your home would be built incorrectly.

- Although the construction phase of a new home may look chaotic, each construction operation is done in a logical sequence and follows a set pattern.

- Some construction activities must be done before others while certain activities can be done whenever the builder wants to do them without affecting your completion date.

- A construction schedule constantly changes due to: weather, logistics, and the availability of labor and materials.

- Whatever you do, **STAY AWAY FROM YOUR HOME DURING THE LAST TWO WEEKS OF CONSTRUCTION!**

- The last two weeks of construction are the messiest and most chaotic of the building process. This period is called "later", because this is when those small items that get put off until "later" have to be done.

- Do not make lists of items that are wrong with your home during the construction process. Your list will contain items that your builder has already seen and will be fixed.

- If any corrections need to be done to your home, they must be done in the correct job sequence. Therefore, they may not be done instantaneously.

Notes

Notes

Chapter 13:
Wrapping It Up

I know that I have asked you to be logical and patient throughout this book; now I am asking you to continue for just a little longer because soon it will pay big dividends.

When your home is almost finished and at the "later" stage, where every task that was put off until later in the construction process has to be done, much more is taking place than during the beginning of your home's construction. Paint is being touched up, the electrician is installing light fixtures, the finish flooring is going in – these activities make the last two weeks of construction the most chaotic time in the building process. Everyone is feverishly working to finish your home so everything will be ready when your furniture arrives.

During this last minute rush to make your home perfect, the most important item to complete is not the cleaning or the paint touchup, but the setting of your closing date. On this date the builder transfers the deed and hands your keys over to you. The closing date usually must be scheduled two to four weeks in advance to allow your closing agent time to prepare your paperwork properly.

So several weeks before your home is complete, your builder or his representative will call you to arrange a convenient closing time. This call seems to be the biggest single source of panic to a new home buyer because they have been by the job several times, (Remember when we told you not to go by the job?), and have seen the horrendous sights of the "later" stage. And after seeing the bathtubs filled with trash and the muddy carpet, someone is now calling them to cheerfully ask when they would like to assume ownership of this beautiful home.

At this point the builder starts receiving panic-stricken phone calls. If I had a ten-dollar bill for every time a client told me, "I'm not closing until everything is finished on my home," I could retire in style. Being frightened and not knowing what to expect, these clients are always surprised when I tell them, "Well, of course you're not going to close before your home is finished. I'm not either. But we do have to set up a time early to make sure we can establish a convenient closing date for us both."

About this time well-meaning friends magically appear and offer constructive criticism like, "My God, they will never get your house done in time!" and, "Who is going to clean this pig sty?" As these friends wail and moan over your home, you can sit back and relax knowing that you do not have to worry about a thing and your home will turn out beautifully. How can you be so sure? Because you picked the right builder, remember? You investigated the builders in your area and picked the best one based on the research techniques we talked about in Chapter 6 - Choosing Your Builder. Your builder knows what he is doing because if he did not, you never would have picked him to build your home in the first place. So sit back, relax, set the closing time, and let your hand-picked qualified builder finish the job you hired him to do.

Besides, if you decide that your home is not complete when the closing date approaches, you have a built-in safety net. Like any other activity in a construction schedule, you can simply postpone the closing until your home is complete. Even though this safety net is available, I can count on one hand the number of times in more than thirty years we have actually postponed a closing and still have fingers left over. Once the homeowners see everything completed right on schedule and realize that their home is absolutely beautiful, they try to speed up the closing instead of postpone it. If you just keep the faith you have in your hand-picked builder, you will find yourself trying to hurry your closing date as well.

After you set your closing date, you will coordinate a time with your builder for the walk-thru on your new home. The walk-thru is a final inspection of your home by you and the builder to make sure that everything is complete and list any items that still need attention. Every builder in the country has their own method of doing walk-thrus; however, there are some basic things you and your builder should do before and during the walk-thru to thoroughly check your home.

When your home is almost ready for final inspection and walk-thru, the main thing to remember is not to get in too big of a hurry to do your walk-thru and make your punchlist. When it comes time for your walk-thru, the home should be at a point we call "substantial completion," which means you and your family could move into your home at this stage and live comfortably. So rather than be concerned about how many walk-thrus you get, (a typical first time buyer question) make sure your home is at substantial completion and do **one (1) complete walk-thru** where you thoroughly check every part of your home. The moment you enter your home at substantial completion:

➢ the appliances are installed and working
➢ the electric meter is set and the electrical power is turned on
➢ the A/C and heat are set and ready to turn on
➢ the water is turned on
➢ the home is thoroughly cleaned
➢ all construction work is done on the home

Some builders I know try to do walk-thrus without even having the water turned on or the electric meter set. This is **COMPLETELY WRONG.** The purpose of a walk-thru is to check the home thoroughly for minor problems before you move in so that they do not become major problems after you move in. If you try to do a walk-thru before the home is ready for inspection, then you will only see the chaos of an incompleted job. You will subject yourself to needless apprehension and fear and create a punch list the size of an average encyclopedia. A punch list this massive is useless because it is so filled with items that are obvious and routine that none of the workers on the job will bother to read it.

For example, if the punch list says there is no handle on the front door, the trim man will notice something this obvious in his sleep. He will put the door handle on whether or not it is on the walk-thru punch list. Therefore, if he sees this item on the list, he will regard the list as useless and throw it away without ever reviewing it, leaving important items in disrepair.

No matter what kind of walk-thru procedure your builder uses, you never ever, <u>ever</u>, **<u>ever</u>**, **EVER** want to do a walk-thru early under any circumstances. **<u>Never do it!</u>** Only once in my long career did I allow a walk-thru early, and I will never, **under <u>any</u> circumstances**, <u>**ever**</u> do it again. It happened about fifteen years ago with a young, first-time buying couple who had to leave town because of a dying relative and wanted an early walk-thru so their home would be finished when they returned. I had serious reservations because their home was at the "later" stage and their electric meter had not even been set, but they begged and pleaded so I acquiesced and granted their request. I explained to both of them extensively that their home was not ready for a walk-thru and therefore we would have more items to complete than normal.

The couple said they understood this, but the next day as the July sun roasted the home with 105 degree heat, they became less and less understanding. With every paint scratch and nick they found, their patience eroded more and more until finally the husband began screaming at the top of his lungs that I was defrauding them and I was not going to get away with it. At that point I simply told them that the walk-thru was over, and we would have to reschedule this inspection. When they returned from their trip several weeks later and saw their home completely done with the A/C set at 75 degrees, the walk-thru went flawlessly, and we all lived happily ever after.

If I were faced with this same situation today, I would ask the couple to attend to their family, have a safe trip, and take care of their home when they returned. If an unexpected emergency like this comes up, the best thing to do is postpone the walk-thru and come back to your home when you are ready. The last thing you ever want to do is jump the gun and do a walk-thru before your home is complete. Because if you do, you are just asking for trouble.

As you prepare for your walk-thru, remember that this is not a period for confrontation; it is a period of explanation. It is a time to inject as much logic and as little fear and panic into the process as possible. With this in mind, do not be surprised when your builder asks you not to bring your kids and relatives to your walk-thru. This is not a family outing nor a time for your friends and relatives to offer "helpful" advice – they will only inject fear and panic into the walk-thru process. Instead, this is the time you and your builder should tour the home alone and do so thoroughly, with no distractions. Inspect every item, learn how each piece of equipment works, and get information on routine maintenance items necessary for proper home upkeep. This is a serious time when you need to focus completely on what you are doing. If you are worried about your child hurting themselves or listening to them yell, then you cannot focus on what is important. Therefore, we always ask parents to leave their children at home so we can concentrate on what needs to be done to make their house a home. Do not be surprised if your builder asks you to please do the same.

In addition to relatives, our policy is to not allow any so-called "expert consultants" to join us during the walk-thru. When you selected your builder, you did your homework and picked a builder who is an expert in his field. He is the expert working for you and you have paid him a substantial amount of money to guarantee that your home is built right. So why pay more money to someone who may not be as skilled as the professional you already have? My estimating teacher at the University of Florida put it best when he said most consultants were like "a drunk at a party who stands in the corner and pees on himself. You don't know what he's doing, he doesn't know what he's doing, but he's getting a warm, fuzzy feeling from it."

I remember the walk-thru we had several years ago which established this company policy. The homeowners, their "consultant", and I were walking through the home and inspecting each room. Whenever I tried to explain something to the homeowners, their "consultant" kept trying to refute everything I said. I do not mind having something pointed out to me if it could be a problem, but this man kept trying to contradict everything I said and kept getting his facts wrong. The final straw came when I was explaining how to operate their heat pump system and warned them not to switch between heat and cool immediately because the heat pump would burn itself up. At this point, the "consultant" puffed out his chest and exclaimed, "On my heat pump, I can switch between heat and cool as much as I want to."

At the end of a three-hour session, the last thing I need is some ignorant buffoon who knows nothing about building questioning me. I did not go to school for four years and work over thirty years in this industry so some know-nothing upstart could make my life more difficult for his amusement. So as you can imagine, after three hours I was quite sick of this "consultant". When he questioned my knowledge of heat pumps, I slowly turned to him and replied, "Yeah, you idiot, you can switch between heat and cool as much as you want. You can also void your warranty and have to buy a completely new heat pump system!"

Due to this and other similar experiences, our company policy is that if the owner brings someone not a party to the contract to the walk-thru, they will not be allowed into the home, and the walk-thru will be rescheduled. Do not be surprised if your builder has a similar policy, because I am sure similar experiences have made this his company policy as well.

As we have said before, every builder may have his own unique routine for doing a walk-thru. Since each routine serves the same basic purpose, inspecting your home, I am going to describe our walk-thru process and ask you to please not become alarmed if your builder does his somewhat differently. Although his method may be different, he still thoroughly inspects your home and makes sure it is completely finished when you and your family move in. So just realize that in this case it is okay to eat your dessert first as long as you also eat your veggies before you leave the table. In the end, you will still get the same nutritional, well-balanced meal.

On the morning of your walk-thru, my superintendent or I turn on all the electrical breakers in the main electrical panel box and make sure all the electrical equipment is working. I then set the A/C or heating system to a comfortable temperature and lock the home. By now, we have changed the lock from our standard construction key, which all our workers use to enter the home, to the privacy key which the owner will be given at closing. This way, nobody except me can enter your home until the appointed walk-thru time. Now that everything is clean and ready for inspection, it is important to keep the home locked so the workmen will not walk on your carpet with muddy boots or scratch the walls or do any other unsightly things before we inspect your new home.

So at this point the home is locked tight, the A/C is running, and we are ready for that afternoon's walk-thru. At the scheduled time I arrive at the home and if the homeowners are not there, I leave the home locked and wait until they arrive. After we have shaken hands, I explain that this walk-thru will be the final inspection of the home. Together we will go through each part of the home room by room and examine each and every part of that room before moving to the next one. This way, nothing is missed because one person is in the foyer while another person is in the bedroom. Electrical outlets, flooring, windows, doors, plumbing, appliances, electrical fixtures, electrical equipment – every item is checked during this final inspection. Anything that needs to be done in each room will get on the builder's punch list and will be conveyed to the painters, plumbers, electricians, and dozens of other workmen that may be involved in correcting the walk-thru items. All of the items you find must be on this to-do list, because if an item does not get written down, it will not be conveyed to the workers, and it will not be corrected.

As we go though the home, I make a list of anything that requires fixing or touching up, beginning with the home interior. We always work from the inside out because the workmen will address the items on the list in the order they are listed. That way, if the workmen still have something to finish after you move in (which is rare but can occasionally happen), they can minimize the inconvenience by completing their work without you having to be home.

Once you cross the threshold and begin your walk-thru, the first thing to do is throw away any lists you made before your walk-thru. Go into the house like you are walking in for the first time and examine everything with a fresh eye. I know some of you will want to make sure the blemishes from before the walk-thru are repaired, but remember that the workmen have been finishing your home for the last two weeks. Chances are that they did most of the stuff on your list, so why waste your time looking at things that are fixed already? Make sure to concentrate on what is actually wrong now, **<u>not</u>** what was wrong three weeks ago.

In all my years, I have found that although we check your home out from top to bottom during the walk-thru, the majority of things wrong with a home at this stage are paint touch-ups. One reason for this is that everything else on the home from electrical to plumbing systems is already completely finished and checked out by this time. The second, and more important reason is that most people only check the paint during a walk-thru and ignore everything else because smudged paint is the easiest item to catch.

As a smart homebuyer, you should realize that **<u>paint</u>** is the **<u>least</u> <u>important</u>** item to check and the easiest to repair in your new home. You will have to look for much more than scratches on the wall of your new home if your walk-thru inspection is going to be worthwhile.

Many times over the years, I have opened the front door and have seen homeowners visibly relax as the walk-thru began. They see the floors have been cleaned, the walls have been painted, and feel the cool A/C blow away their fears and anxieties. Now they realize that everything will indeed be okay and anything that remains to be done before closing is so minor that they could live in the home right now, even if the walk-thru process were never completed. With their construction worries over, the homeowners are now free to worry about other important matters such as hiring furniture movers and making sure everyone knows their new address.

Sometimes an especially nervous client will ask, "How do we know everything will be done before the closing?" Well, the easiest way is to come back and look around before your closing. But just remember that your next visit after your walk-thru should never turn into a second walk-thru. It is very possible that on this visit you will find paint scratches that you did not see on the walk-thru (in fact I would be **very** surprised if you did not) because of different lighting conditions or because of a workman who scratched the wall accidentally with one of his tools. So when you see extra scratches, realize that it is very easy to grab some paint from the touch-up kit left in the garage and paint them. Most people make more dents and scratches moving their furniture into their home than were there before the walk-thru. That is why we leave a touch-up kit with extra paint in the garage. While you are painting over the scratches from your furniture, just repaint those few places on the wall we missed during the walk-thru and your home will look absolutely beautiful.

Normally, we set up our walk-thrus for two or three days before the closing to give ourselves some time to schedule any necessary tradesmen to come back without putting them under extreme pressure. Once in a while, you may encounter the unexpected such as a faulty circuit breaker or a leak under the kitchen sink that requires a minor callback. Giving the tradesmen an extra day or two of leeway will allow them to get any necessary parts if they are out of stock and make sure your home is done right before you and your family move in.

After the walk-thru, your sales representative will give you the final closing information you need, including the final amounts that remain to be paid in your contract (since you have a detailed, written contract there should be no surprises at this point), and set up the last minute items for your closing. At the closing, your closing agent will explain each of the closing documents in detail and have you sign them. Then you will give the closing agent your money, and the home will officially be yours. The builder will then give you some instruction on maintaining your home, review your home warranties, and tell you who to contact for any minor callback work. Next he will present you with the keys and your closing package[i] and your dream home will finally be yours.

Now you need to remember that like your automobile or any large purchase, your home may have some minor items that need attention after you move in. Most homes come with a standard one year warranty plus some kind of extended limited warranty on certain parts of your home. So during your warranty period, if you notice a problem that is not a drastic emergency, write it down. When your list gets to be about three or four items long, call your builder and set up a time for him to send someone out to take care of them. This way you can minimize the number of trips for your builder and the amount of time you need to arrange to be home to meet him.

In case you have a problem with large items such as plumbing, A/C or appliances, you should receive a list of subcontractors who handle these items, along with their phone numbers, in your closing package. These subcontractors have repairmen on call who can get to your home and repair your problems quickly and easily. Whether it is during normal working hours or on weekends, the subcontractors like for you to call them directly so they can set up a time with you and avoid playing "telephone tag" or finding no one home and wasting a trip. With this subcontractor list, you can set up an appointment directly and avoid the wasted trips that can so frustrate you, your builder, and your repairman.

When a repairman shows up at your door, please remember that even though your A/C unit just blew up and it is 100 degrees outside, it is not his fault. Please do not take your anger out on him because he is most likely not the same person who installed your unit in the first place. And even if he is the same man who installed your unit, equipment such as A/C units are very complicated and have many components that can break down without warning and are impossible to detect beforehand. So it will serve you well to be as kind as you can because they are people too, and they respond to kindness the same way you would. If you are kind to them, you will get a better job, and it will be easier to get them to come out the next time if you are unfortunate enough to need their services again.

A perfect example of this happened last summer when I remodeled the kitchen in my home. While the men were working, my wife made them lunch and served them refreshments until they thought they were going to burst. When they were through installing our new kitchen cabinets, it was the best job I have seen in over thirty years. I mean my kitchen cabinets look better than custom kitchens I have seen in homes three to five times the cost of mine. And the reason is not because these men were so much better than other cabinet installers; rather, they were made to feel very comfortable while they worked. My wife made sure they realized she appreciated the hard work they were doing for her, and in return they did above and beyond what the job required. So when you have workmen come by your home to do something for you, serve them a cup of coffee or a cold drink or even some food when lunchtime rolls around. When you make the workmen feel appreciated in this way, you will get a better job.

Being kind not only works for the men who repair your home once you have moved in, but also works for the tradesmen you see on the jobsite during your home's construction. Word gets around fast when a difficult homeowner constantly complains and chews out men on the jobsite, especially if they are complaining about things the workmen did not cause and/or cannot rectify. So when you visit your home during construction, please treat the men you meet with kindness and respect. A few encouraging words such as, "You guys are doing a nice job," or "We sure appreciate you guys doing such a good job," will make the workmen want to do their job right for you. I have even had some homeowners go so far as to bring soft drinks out to the men on the job to show their appreciation. And though it may only be a coincidence, those are the homes that had the least number of problems to correct once we did the walk-thru and the final wrap-up of the home.

<u>Highlights of Chapter 13</u>

- The closing date for your home must be scheduled two to four weeks prior to your home's completion so your closing agent has time to prepare the proper paperwork.

- You can rest assured that your home will turn out beautifully, no matter what your friends tell you, because you researched your builder carefully and picked the right one.

- If your home is not complete when the closing date approaches, you can simply postpone the closing until your home is complete.

- The walk-thru is a final inspection of your home by you and your builder to make sure everything is complete and list any items that need attention.

- The home should be at the stage of "substantial completion" at the time of the walk-thru. This means your family could move into the home at this stage and live comfortably.

- Rather than be concerned about how many walk-thrus you get, make sure your home is at substantial completion and do one (1) complete walk-thru.

- **Never do a walk-thru early, under any circumstances.**

- The walk-thru is not a period for confrontation with your builder; rather, it is a time to inject as much logic and as little fear and panic into the process as possible.

- Your builder will most likely ask you not to bring your children or other family members to the walk-thru so that you will not be distracted by them.

- Your builder will most likely also ask you not to bring any so-called "expert consultants" to the walk-thru. Your builder is already an expert in building and you have paid him a substantial amount to make sure your home is built correctly. So why pay for something you already have?

- Any person who comes to the walk-thru and is not a party to the contract will most likely not be allowed in the home. Your walk-thru is a time for you and your builder to inspect your home with no distractions.

- During the walk-thru, you and your builder need to go through each part of the home room by room and examine every part of that room before moving on to the next one.

- Every walk-thru item that needs to be corrected must be written down on the builder's punch list, or these items will not be corrected.

- Do not take any lists with you to the walk-thru. Go through your home like you are walking in for the first time and examine everything with a fresh eye.

- Realize that paint scratches and blemishes are the **least important** item to check during a walk-thru and the easiest items for the workmen to repair.

- To make sure everything is done before closing, come back and look around before you close. Just make sure that this visit does not turn into a second walk-thru.

- After the walk-thru, your sales representative will give you the final closing information you need, including the final amounts that remain to be paid in your contract and set up the last minute items for your closing.

- At the closing, your closing agent will explain each of the closing documents in detail and have you sign them. Then you will give the closing agent your money, and the home will officially be yours.

- The builder will next give you some instruction on maintaining your home, review your home warranties, and tell you who to contact for any minor callback work. Then he will present you with the keys and your closing package and your dream home will finally be yours.

- During your warranty period, if you notice a problem that is not an emergency, write it down. When your list is about three items long, call your builder and set up a time for someone to come out to take care of them. This way you can minimize the number of repair trips and the amount of time you need to arrange to be home and meet the repairman.

- If you have problems with plumbing or other large items, call the subcontractors in charge of those items directly to set up a repair time.

- When the repairman comes to your door, please do not take your anger out on him. Your problem is not his fault and he will do a better job if you are kind to him.

- Being kind not only works for the men who repair your home once you have moved in, but also works for the tradesmen you see on the jobsite during your home's construction.

[i] closing package– given to you at closing, it contains information on warranties, maintenance, and a list of subcontractors who performed work on your home

Notes

Notes

Chapter 14:
Final Thoughts

Years ago when I was in college, one of my instructors observed that the public has a very limited view of builders because the information they receive is always at least secondhand. Builders never get to tell their side of any story because builders never write and writers never build. So all the tales of new home victims you hear passed around and generously embellished are reported by people who do not understand the process. Then the next thing you know, our reputations as builders are tarnished just a little bit more. In fact, I have even seen some really good movie plots in which the villains of choice are greedy developers and dishonest builders trying to cheat the public out of their hard-earned housing dollar.

People who advocate zero growth and do not want to see anything built anywhere, or **BANANA's** (**B**uild **A**bsolutely **N**othing **A**nywhere **N**ear **A**nything) have characterized builders and developers as greedy, dishonest usurpers of the land who seek to destroy nature and devastate the ecology for a few pieces of silver. Now these pseudo-ecologists (not to be confused with people who are genuinely and deeply concerned about the environment) only use these thoughts and concerns to stop future building for their own ends. They have already purchased their portion of the American dream and now want to deny those who come after them the same opportunity to purchase a home. I have often been amused when these hypocritical buyers insist we take down all the trees when their home is built only to show up to protest a new proposed community for "ecological concerns." Their concern is not to protect the environment; their only concern is to make sure that nobody else builds in their area.

This scenario happens and it happens often. But the truth is that reasonable developers and builders understand better than most the importance of preserving the environment. They know that nature is the greatest of all designers, and the esthetics of nature are irreplaceable. They also know that good ecological design sells far better and it sells for a higher price. Therefore, no builder or developer in his right mind would wantonly destroy the landscape that nature has done such a beautiful job providing for his future clients.

Whenever a new development comes online that happens to be without any trees, everyone automatically assumes that the builder/developer is at fault. He is the one who came in and took out all the trees and converted the once lush, tropical forest into a barren wasteland for his own selfish gain. But the truth is that in most cases the cause lies not with the developer but with the farmer who converted the land to improved pasture by taking out all the trees years or even decades before. Now, the city has built out to the pastureland, the farmer sells out, and the developer must spend thousands of dollars to bring in trees and reverse the process. And while the developers are working to repair the ecological damage, an unknowing public places the blame on the builders and developers who are trying to fix a problem they had no part in creating.

In addition to greedy developers, you often hear stories of dishonest builders who cheat their buyers and violate the building codes in various ways. One of the most common stories I have heard is of builders pulling reinforcing steel out of concrete foundations before they are poured or substituting lower grade lumber and other materials for structural uses. Whenever you have a natural disaster, the first thing that comes to mind as victims are pulled from the wreckage of their homes is that the builders bribed local building inspectors to reap huge rewards by violating building codes.

Well, I can truthfully say that in more than thirty years I have never seen any such thing happen. I suspect that if it does happen, it is very, very infrequent, if for no other reason than **bribes and kickbacks** to local building inspectors are absolutely **not economically feasible**. In a case like this, crime simply does not pay. For example, lets take the myth that a builder commonly removes all of the reinforcing steel out of his foundations to save money. An average home in Florida requires thirty to forty bars of reinforcing steel in its concrete foundation. Now these reinforcing bars cost approximately $3.50 each. So if a builder removed every piece of reinforcing steel from the average home he would save approximately $140. And out of this $140 he would still have to pay the labor cost for the removal of the bars and the cost of transporting the bars to the next job for reuse. So if the cost of removal labor and transportation came to only $40 (a very low figure, by the way) that would mean that the builder and the inspector could **split a whopping $100** between themselves.

It is inconceivable that anyone would violate the law, jeopardize more than $100,000 worth of new construction, and risk having their building license revoked for a mere $50 in savings. And yet, this is one of the most enduring myths about builders in our society today. The same logic holds true for using cheaper grades of structural lumber and other such materials. If all of these alleged tales of horror were true, I doubt very seriously that these accumulated horrors could save more than **$500 on an ordinary house**. They say that every man has his price, but few would risk breaking the law, having their professional license revoked, and having irreparable damage done to their reputation for such a small reward.

The truth is most problems that occur with construction stem not from dishonesty but from ignorance of construction practices. Unqualified supervisors who lack the training and experience to manage their work are being hired by builders who themselves lack the technical knowledge of construction to know bad work when they see it. The last several decades have seen a severe shortage of qualified personnel develop in all of the construction trades. Part of the reason for this is the public no longer has the respect for tradesmen that it once had. There was a time when tradesmen were proud of their skills as good masons or carpenters, but today that is not the case. Because of this lack of respect, even the very high wage rates we have today are not attracting the qualified people to the trades that we once had. From this shortage of qualified tradespeople comes a dependence on a less trained and less experienced labor pool. This means that today's supervisory personnel need to be more skilled, not less skilled, in order to ensure that their work is done properly. However, in many cases, a qualified supervisor simply is not available. This is why you need to be more concerned about the experience and training of your builder and his supervisors than about whether your builder is going to try to cheat you out of a quality home. This is also why we have been telling you again and again that **selecting a qualified builder** at the beginning of the designing and building process **is vital**.

As we wrote this book, my son and I tried to present the designing and building process as a logical sequence of events. Now the order in which the events were presented in this book is not necessarily the order you must follow. Whether you choose your home design or your builder or your homesite first does not matter. What matters is that you make well-informed choices about your new home during each step in the process.

The designing and building experience is filled with wonder and delight. As many times as I have seen it happen over the years, I am still spellbound every time I create a great building design on paper and turn it into a home my clients will enjoy for years to come. Seeing your children grow with each birthday party, watching their eyes light up as they open their Christmas gifts, eating Thanksgiving dinner with loved ones – these are only a few of the many happy memories you and your family will experience in the home your dreams built.

These rewards make any risk you encounter in the process well worth it. Just do not let the risks deter you from building the home of your dreams. Do not be afraid of the process; be very **careful** in the process. Although emotion plays a big part in designing and building, make sure you inject logic and common sense every step of the way and make good, sound decisions. If you allow the advice we have offered in this book to guide you through the process, you should be ready to make the right decisions for yourself and your family.

Now that we have been through the design and building process on paper, I hope we have given you some insight into the design and construction of your new home. We have shared some stories, all of which are true (believe it or not), and I hope we have dispelled some of the misconceptions and alleviated some of the anxiety as you venture forth to design and build your dream home.

Here's wishing that like all good dreams, your home building experience ends happily for you and your family.

Notes

Appendix A:
Glossary of Mortgage Terms

<u>Glossary of Mortgage Loan Terms</u>[i]

Adjustable-Rate Mortgage (ARM) – A home loan with an interest rate and monthly mortgage payment which may change during pre-determined adjustment periods, usually once or twice a year. An ARM may also be referred to as a variable-rate mortgage or an adjustable mortgage loan (AML).

Adjustment Period – Also known as an adjustment interval, it is the time during which the loan rate and/or the monthly payment may change on an adjustable-rate mortgage, usually one or more times a year.

Amortization – A scheduled repayment of a loan in which both the interest and principal of the loan are paid over a specified period of time, usually fifteen or thirty years.

Annual Percentage Rate (APR) – The cost of a mortgage expressed as a yearly rate. This rate is usually higher than the percentage rate stated on the mortgage because it incorporates points and other fees. The APR gives homebuyers a way to compare different mortgage types based on the cost of each loan per year.

Appraisal – An estimate of the value of your home, or the act of arriving at an estimated value for your home by a licensed appraiser.

Appraised Value – An opinion of the value of your home reached by an appraiser after they have finished an appraisal of your home.

Appreciation – An increase in the value of real estate.

Assessed Value – The value placed upon your home for taxation by the local government.

Assessment – see assessed value.

Assumption of Mortgage – When a homebuyer takes over the payments on a home mortgage and becomes the primary borrower or mortgagor on the loan. This may allow the homebuyer to avoid the hassle of applying for a new loan on the property.

Balance Sheet – Needed by those who are self-employed, this document lists the assets and liabilities of a business. These are used to determine the profitability of a business.

Balloon Mortgage – A mortgage with monthly payments that pay part of the principal and interest on a loan but do not fully pay off the loan. The remainder of the loan is then due in a lump sum on a specified date.

Balloon Payment – The unpaid principal amount of a balloon mortgage which is due in one lump sum on a specified date.

Basis Point – 1/100% or one one-hundredth of one percent. This value is used to describe a change in yield for mortgages.

Binder – A deposit on a home by the homebuyer which secures the right to purchase the home at the terms agreed upon when the binder is given to the seller.

Borrower – Person who borrows money for a home and agrees to pay it back by signing a mortgage with a bank or lending institution.

Cap – Also called a rate cap, it is a limit on how much the interest rate or monthly payment may change on an ARM mortgage.

Certificate of Eligibility – Document provided by the Veterans' Administration that details the amount of the mortgage loan that the VA will guarantee for the homeowner. The VA will guarantee loans for active, reserve, and retired military personnel as well as their widowed spouses.

Closing – The act of transferring the ownership or title on real estate from one person or organization to another. During the closing the deed changes hands, mortgage notes are signed, and money changes hands. In most states, a closing is handled by a settlement agent, often an attorney.

Closing Agent – An attorney or legal professional who takes care of the paperwork connected with closing and manages the closing process.

Closing Costs – Expenses incurred during the buying and selling of real estate. These include items such as title fees, loan fees, appraisal fees, etc. The closing costs are usually about three to six percent of the mortgage amount.

Co-Borrower – A second borrower who co-signs a home mortgage with the borrower. The co-borrower's income, assets, and debts are combined with the borrower's in qualifying for the loan. The co-borrower's name also appears on the mortgage, note and deed.

Collateral – An item promised as security for a debt which may be taken by the person loaning money if the debt is not repaid. For example, your home is given as collateral to a lender when a mortgage is taken out on your property. If you do not pay your mortgage, the lender may foreclose and resell your home to cover your mortgage debt.

Commitment – A written agreement between a borrower and a lender to lend money on a certain date under specified conditions.

Comparables – Also known as "comps", these are homes or pieces of property that have sold recently and are similar to the property being considered for purchase. "Comps" are usually homes similar in size, general location, character, and construction type.

Comps – See comparables.

Conventional Mortgage – A mortgage made by a lender without the backing of governmental programs such as the Department of Veterans' Affairs (VA) or the Federal Housing Administration (FHA).

Conveyance – A document, such as a deed or mortgage, that is used to legalize the transfer of ownership on property from one person or group to another.

Conversion Option – The option to convert an ARM loan to a fixed-rate loan for the remainder of the term. This option is usually available each year on the anniversary date of the loan.

Credit Report – A report documenting a borrower's current credit status and credit history. This report is used to help determine whether a borrower will repay their mortgage.

DD214 form – A document issued at the time a service member leaves the military which states information such as: length of service, circumstances of discharge, and location of active duty stations, just to name a few.

Deed – A document which records the ownership of real property changing hands from one person or party to another.

Deed Of Trust – A deed that conveys the title on property to a neutral third party who holds the deed in trust until the mortgage is satisfied. Once it is satisfied, the deed is given to the person who paid off the mortgage. This deed is not used in all states.

Default – A breach of the mortgage, caused by the homeowner not making their monthly mortgage payments.

Delinquency – When the borrower does not pay his debts when they become due.

Earnest Money – see binder.

Equity – The difference between the value of your property and the principal balance due on your mortgage.

Escrow – The temporary holding by a third and neutral party of deposited funds until the terms in an agreement, specifying when the money can be released, are completed. These funds are normally a portion of a homeowner's monthly mortgage payment which the lender uses to pay certain items, such as property taxes and hazard insurance, as they become due.

Fannie Mae – The Federal National Mortgage Association (FNMA), a corporation created by Congress to purchase and sell conventional residential mortgages as well as those insured by FHA or VA. Corporations such as FNMA allow mortgage holding institutions to sell your mortgage to other institutions for a profit. When your mortgage is sold, you then pay your same mortgage payments to whoever bought your mortgage, instead of who issued your mortgage to you.

FHA Mortgage – A mortgage loan with a low down payment insured by the Federal Housing Administration (FHA). The FHA protects the lender against any loss in case the borrower defaults on their loan.

First Mortgage – A mortgage which creates a primary lien against your home. This means that if your home is foreclosed on, the organization which holds the first mortgage would be paid first and all others who are owed money would be paid afterward.

Fixed-Rate Mortgage – A loan whose interest rate remains the same for the life of the mortgage.

Freddie Mac – Federal Home Loan Mortgage Corporation (FHLMC), a semi-governmental agency that purchases conventional mortgage loans from insured depository institutions, such as banks, and HUD approved mortgage lenders.

Gift Letter – A letter from an individual(s) giving funds stating that the funds are a gift and do not have to be repaid by the borrower.

Ginnie Mae – Government National Mortgage Association (GNMA). Originally part of FNMA, it was splintered off in 1968 by an act of Congress and assumed responsibility for special assistance loan programs or programs to assist lower-income families who otherwise could not qualify for a new home mortgage.

Good-Faith Estimates – A requirement that lenders give borrowers an estimate of costs the borrower will have to pay to close on their home. A good-faith estimate must be provided by the lender within three business days of having received a signed loan application.

Graduated Payment Mortgage Loan (GPM) – A flexible-payment mortgage loan where payments increase for a certain amount of time and then become level and fixed.

Hazard Insurance – Insurance coverage providing compensation to the person or group insured in case of a loss or damage to property. The insurance covers damage to the structure of the home only, not damage to the home's contents.

Homeowners' Insurance – Similar to hazard insurance except that this insurance covers both the structure of the home and the home's contents.

Housing and Urban Development (HUD) – Established in 1965, HUD is a government agency responsible for implementing and governing urban development and government housing programs. Their programs include: community planning and development, equal housing opportunity, mortgage credit (FHA), housing production, research, and technology.

Impound Account – See escrow account.

Index – A base for determining the interest rate for an adjustable rate mortgage (ARM). Indexes generally reflect the prevailing market conditions of the time.

Initial Payment Rate – The interest rate established by "buying down" the interest rate whereby the homeowner or the builder pays more money up front to bring the interest rate down below market value and thereby lower their monthly mortgage payment.

Insured Loan – A mortgage loan insured by private mortgage insurance (PMI), the VA, or FHA.

Interest – A fee paid by someone who borrows money for the right to use those funds. Interest is usually expressed as a percentage per year of the total amount borrowed.

Jumbo Loans – A mortgage loan larger than the limits set by Fannie Mae and Freddie Mac. (Any loan greater than $240,000 is considered a jumbo loan as of January 1, 1999). Jumbo loans usually have a higher interest rate because they cannot be funded by these two agencies.

Lender – An individual or company which provides money to borrowers, who want to buy a home in the form of a mortgage. The borrower agrees to repay the lender the amount he borrowed plus interest. If the borrower does not pay, the lender assumes ownership of the home, and the borrower must vacate the premises.

Lien – A legal claim against a property for a debt that is as yet unpaid, such as a mortgage.

Loan-to-Value (LTV) – The ratio of the amount of the mortgage to the appraised value or the sales price, whichever is lower. An LTV is used by lenders to determine how much to loan on a new home.

Lock-In Period – The time during the processing of the home loan in which a certain interest rate is guaranteed to the borrower before closing.

Margin – The constant amount, or percentage, that a lender adds to an index to come up with an interest rate for an adjustable rate mortgage (ARM) at the time of each adjustment.

Market Value – The current estimated value of a home or property that someone will pay if it were being bought today. This, like all estimates, is what one of my college professors used to define as a "scientific wild-ass guess."

Mortgage – A formal legal document pledging that the property will be used as collateral for a loan that the homeowner will repay at a future date.

Mortgage Broker – A person who helps homebuyers find affordable mortgages. The broker charges a fee only when the homebuyer closes on a loan.

Mortgage Commitment – A written document signed by the lender in which he agrees to make a mortgage loan on a specific property and explains the loan amount, conditions, and length of time the commitment is valid.

Mortgage Insurance – Insurance obtained from a privately owned company that provides the lender with protection against financial loss in the event the borrower defaults on the mortgage.

Mortgagee – The lender.

Mortgagor – The borrower or homeowner.

Negative Amortization – An increase in the principal balance for a mortgage loan which occurs when the monthly payment is not high enough to pay the interest due for the loan that month and the unpaid balance is added to the principal amount due from the borrower.

Non-Conforming Loan – A loan which does not meet the requirements of Freddie Mac or Fannie Mae loan guidelines. A jumbo loan or a loan for someone having less than perfect credit would be examples of this kind of loan.

Note – A formal document signed by the borrower stating that he is in debt to the lender and that he agrees to repay the debt to the lender.

Note Rate – The actual interest rate being paid under the terms of a mortgage versus the APR.

Origination – The process in which a mortgage lender obtains an application, gets approval by a lending institution who agrees to loan the money for the mortgage, and issues a loan commitment on a mortgage loan secured by real property.

Origination Fee – The fee mortgage lenders charge homebuyers for preparing loan documents, making credit checks, etc. This fee is usually computed as a percentage of the face value of the mortgage loan and is usually paid by the homebuyer or the builder, depending on the contract terms.

P&L – Required only for people who are self-employed, it is a profit and loss statement which details the money flowing in and out of a business.

PITI – An acronym for the items that are included in a monthly payment: **P**rincipal, **I**nterest, **T**axes, & **I**nsurances.

Point – Also known as discount points, an amount equal to one percent of the mortgage loan amount. For example, if the builder offers you two points on a $100,000 home, that means that he will pay $2,000 or 2% of $100,000 and "buydown" the interest rate so you will pay a lower interest rate on the $100,000 loan.

Preapproval – Also known as credit and income approval, this is where credit and income information given during the prequalifying process has been verified by your credit report, bank statements, pay stubs, W-2 forms, and tax returns.

Prepaids – Expenses that must be paid to create an escrow account or adjust the seller's existing escrow account. These expenses may include: hazard insurance, taxes, PMI, and any special property assessments.

Prequalifying – Process by which the loan officer uses debt and income information to determine the amount you can spend on a new home. Because none of the information you give to your loan officer at this time has been verified, your prequalifying amount is the amount you can spend on a new home if all of the information you have provided is verified and accurate.

Prepayment – A clause in a mortgage which permits the borrower to make mortgage payments before those payments are due. Not all mortgages allow a borrower to do this.

Principal – The amount of money borrowed, excluding interest and other fees.

Private Mortgage Insurance (PMI) – see mortgage insurance.

Purchase & Sale Contract – The written contract whereby the seller agrees to sell the home and the buyer agrees to buy the home for a specified amount.

Recording Fee – A charge for recording the transfer of property (deed) from one owner to another. This fee is paid to the city, county, or other branch of government where it is required in your area.

Sales Contract – Contract between the seller and buyer which explains in great detail: what the purchase price includes, guarantees on the property, when the buyer can close and move in, closing costs, and what will occur if the contract is not fulfilled or if the buyer cannot procure a mortgage at the agreed upon terms.

Second Mortgage – Borrowing money against a home that already has a mortgage on it and not paying off the original mortgage.

Security – Real property given, deposited or pledged as a guarantee that a debt will be repaid.

Start Rate – see initial payment rate.

Statement of Service – Letter prepared by the commander of the squadron for active duty personnel stating employment information such as: income, pay entry base date, reenlistment date, et cetera.

Term – The length of time in which a loan must be paid.

Title – The documentary evidence of property ownership. In real estate, this is the title deed which specifies who has legal right to ownership of the property.

Title Binder – A written report issued by the title insurance company that gives a listing of all the defects in the title, the names of the current and proposed owners, and any liens against the property it defines. The binder can only be used as title insurance temporarily and must be replaced by a permanent title insurance policy.

Title Insurance – Insurance written by the title company to protect the lender or the owner against loss of money in case former liens or former owners of the property try to lay claim to the property to satisfy debts that were not paid previously. The cost of this insurance depends on the value of the property and is usually paid for by the borrower.

Title Search – A search of public records, court decisions, and current laws to discover the facts about the ownership of and liens against a property.

Truth-In-Lending – A federal law which requires the lender to tell the borrower, in writing, the Annual Percentage Rate they will pay on their loan, their credit terms and conditions, and other mortgage charges they are subject to within three days of loan application.

Underwriting – Analyzing the risk to the lender if they lend money to a person applying for a mortgage on a new home. In this analysis, factors such as: credit, employment history, assets, the value of the property the mortgage is applied for, and other factors are taken into account and used to judge whether the loan should be made and what interest rate the person should have to pay.

VA Mortgage – A low down payment loan for a home guaranteed by the Veterans' Administration (VA) and offered to eligible veterans.

Veterans' Administration (VA) – A federal agency that insures mortgage loans for honorably discharged veterans and their spouses.

i   "Step by Step Guide to Home Financing", North American Mortgage Company brochure

ii   "1999-2000 Home Finance Guide – A comprehensive resource for real estate professionals", Charter One Mortgage, A Subsidiary of Charter One Bank, F.S.B. brochure

iii   "Home Buyers Survival Dictionary", Bonded Builders Home Warranty Association brochure

Notes

Appendix B:
Glossary/Index of Construction Terms

Glossary/Index of Construction Terms

A&D Loan – Acquisition and Development loan, this is money a developer borrows from the bank to develop land. He later repays this loan with interest from the sale of homesites in his development. (16, 33)

Aerator – A filtering system for an individual water supply system (well) which filters out many impurities found in ground water. (153)

Allowances – A specific amount of money set aside in a construction contract for certain categories of construction items (i.e. lighting fixtures, appliances, landscaping, etc) (18, 33, 61 - 66)

Anchor Bolts – Steel bolts which are embedded in concrete to fasten exterior and bearing walls to the foundation. (55)

Angle Iron – An L-shaped piece of steel commonly used to support brick or stone above window and door openings. (55)

Appraiser – A licensed professional who appraises or evaluates the value of your home based on the sales price of homes comparable to yours or "comps" in your area. (20, 21)

Architect – A professional who designs and/or supervises the construction of a building. A registered architect is licensed by the state in which they practice, usually holds a college degree in architectural design, and must go through an apprenticeship program prior to taking the state board exams. (45, 65)

Asphalt Shingles – One of the most common types of roof covering materials, these shingles are composed of asphalt treated membranes and a granular wearing surface that is exposed to the weather and resists the elements. Available in several different styles, thicknesses, and colors. (57, 176)

Assessments – Amounts charged by local government to extend services such as water and sewer into an area where they were previously unavailable. (61)

Bar Chairs – Small, wire supports which are used to hold reinforcing steel above the level of the bottom of the footing or form where they are located. (55)

Benchmark – Point of known elevation used to set other elevations such as a home's finish floor level in a subdivision.

Bid System – A method of choosing a contractor based upon awarding the job to the contractor who submits the lowest price for doing the work. (39)

Brick Veneer – A type of wall construction wherein the brick is applied to the exterior of a wall system, usually wood frame construction. This gives a durable, low maintenance exterior while providing an interior wall which is dry, straight, and capable of being heavily insulated for energy savings. The air space between the exterior brick and interior wall covering provides extra insulation and prevents water infiltration due to seepage and condensation. (62)

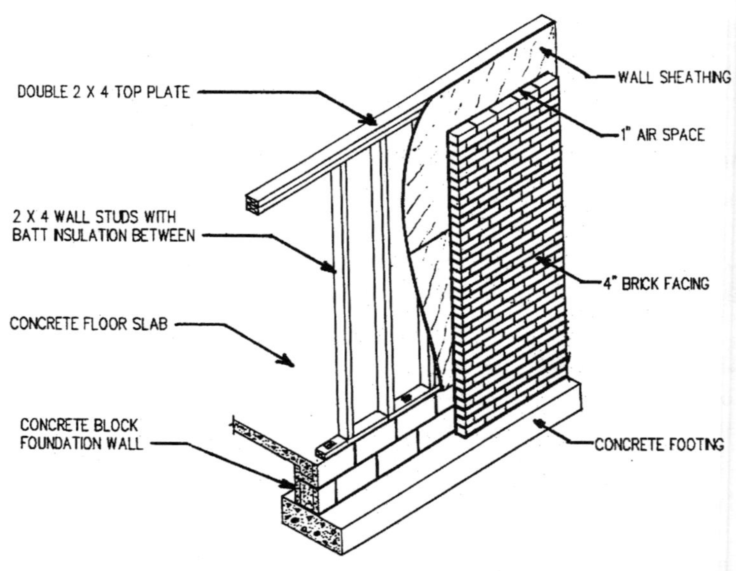

ISOMETRIC OF BRICK VENEER WALL

Builders' Risk Insurance – Insurance that covers damage to the structure while it is under construction from: fire, wind damage, storms, etc. Note that this insurance usually covers only the structure, and not the contents inside the building. Therefore, if you move in before closing, this insurance would not cover any damage to your furniture or personal property. (61)

Building Codes – Governmental regulations which state guidelines on how buildings are to be built in a certain area. Usually each region of the country has their own code governing the building in their area or they may use a standardized code with specific amendments for their area. (27)

Building Inspector – An individual employed by local government agencies or a supervising architect whose job is to check the construction at specific intervals to ensure that the home complies with local codes. (26 - 27)

Building Restriction Line – Line showing the closest point where a structure can be build from a property line.

Certificate of Elevation – A form provided by the Federal Emergency Management Agency (FEMA) which a surveyor fills out to certify the finish floor elevation of a home. These forms are required for those homes located in a flood plain, but usually are not required for other homes.

Closing Costs – The costs associated with closing on a home. These costs may include a final survey of your home, a fee for appraising the value of your home, closing attorneys' fees, title insurance, taxes, your mortgage company's fees, and other such costs. Both construction loans and permanent loans have closing costs associated with them; therefore, it is important that your contract states who pays these costs. (208, 213)

Codes – A body of rules and regulations that govern construction and land use. Examples would include: building, electrical, zoning, handicap, energy, etc. (27)

Comps – During an appraisal, comps, or comparables, are homes which have recently sold in the local area and are similar to the home being appraised. The sales price of the comps and its features are compared to the home being appraised and used by an appraiser to establish a value. (20, 21)

Concrete Slab – A horizontal area of concrete several inches thick which can be used as a floor in a home, garage or basement, a patio, a sidewalk, or driveway, etc. Slabs may be reinforced with steel to accommodate loads which are placed on them. (61, 62)

Construction Loan – A short term mortgage, the proceeds of which are used for the construction of a building and improvements, and/or site acquisition. Funds are paid to the contractor at various stages of construction and the entire loan is paid off at the closing by the permanent loan when title to the property is transferred to the owner. (109, 118, 213, 214, 229)

Control Joints – A joint that is either formed or sawed in a concrete slab that serves to weaken the slab and cause any cracking that may occur to happen in that joint instead of occurring randomly in the slab. (55)

Control Points – Permanent reference points such as intersections of street lines and points where curves begin and end, that are set within a subdivision to facilitate the setting of other property lines.

Cost Overrun – The phenomenon of going over a predetermined budget during the planning or construction stages of a new building or project. (196)

Cost-Plus Contract - Contract whereby the owner pays to the contractor the total cost of all work done on the home. In addition, the owner pays either a fixed fee or a percentage of the cost, which represents the contractor's general overhead costs and profit.

Course – A single horizontal section or row of brick or concrete block work.

Culvert – A drainage structure located underground which usually consists of a large metal or concrete pipe. This underground pipe allows water to be carried or drained from one area to another. (152)

Curb – The edge of pavement, used to channel stormwater.

Curve Data – Table that shows data for each curve line in the subdivision. Shows such things as radius, length, chord bearing and distance, et cetera.

Dedicated Circuit – An electrical circuit used or dedicated to a single purpose such as a computer, microwave oven, etc. (19, 33)

Deed Restrictions – Rules governing the use of a specific parcel of land such as subdivision. These rules are placed on the subdivision by the developer and usually specify such things as: minimum square footage of homes, type of exterior construction, fencing, swimming pools, etc. (26)

Door Jamb – The finished frame in which a door is installed. This frame is most commonly constructed of wood or metal. (57)

Dozing Tank – An ordinary septic tank with a large pump inside. This pump is connected to a float inside the tank which will activate the pump when the tank fills to a certain level. The sewage is then transported to the regular septic tank where gravity pushes the sewage downhill to the drain field for treatment. (156)

Draftsman – A person who draws house plans and/or designs homes. A draftsman is not required to have a license or a college degree and may have varying amounts of experience, thus making it difficult to establish the competence of a draftsman. (45, 65)

Drainage Plan – Plan showing the elevations of property corners and the direction of the sloping ground based on contour lines and spot elevations of curves and road centerlines. Shows the required minimum elevations for the homes on each lot and the direction stormwater will drain off of each lot and how it will be conveyed out of the subdivision (pipe, ditch, etc). (135, 136)

Draw – A partial payment on a contract, usually paid at predetermined intervals of construction established by a draw schedule. (212)

Easement – A portion of a piece of land that can be used by other parties for specific purposes. They give the easement owner the right to do work within the easement and do not allow the property owner to build permanent structures in the easement area. (132, 133)

Electrical Service – The point at which electricity comes into your home. The service is rated in units called amps. The higher the amp rating, the more electrical items you can run off that panel without causing a problem. (19, 33)

Engineer – Also known as a Professional Engineer (PE), this professional designs and/or supervises the construction of underground utilities as well as designing the support structures which ensure the structural integrity of a building. A professional engineer (PE) is licensed by the state in which they practice, usually holds a college degree in engineering, and must go through an apprenticeship program prior to taking the state board exams. (151)

Existing Grade – The level of the ground before any work in done on a home. (155)

Expansion joint – A joint in a concrete slab or other continuous surfaces that is usually filled with a material that will compress or expand as the surface on each side moves due to thermal expansion and contraction. (55)

Felt – A paper-like fiber impregnated with asphalt and used as an underlayment for shingles or behind brick veneer walls to provide a water-proof membrane over decking or wall sheathing. (56)

Finger joint – Wood trim made up of short pieces that are glued together with a specially milled joint for additional strength. These joints are completely undetectable when the wood is painted. (57)

Finish Flooring – The floor covering which is the wearing surface on a floor. This flooring may be: carpet, sheet vinyl, tile, wood, etc. It may be applied over a concrete slab or a subflooring material. (41, 158))

Finish Grade – Level of the ground after the home is built. (155)

Flashing – A thin sheet of metal, plastic, or similar material used to prevent water infiltration where two surfaces of a house intersect and provide a crack where water might penetrate. It may also be used where a surface is penetrated by an opening such as where a pipe penetrates a roof. (57)

Flood Plain – Area that will be flooded by at least one storm per every so many years, usually 50 or 100 year intervals.

Floor Break – Line drawn on a floor plan or home slab to indicate the transition point between one type of finish flooring and another. For example, a floor break could delineate between ceramic tile and carpet. (48)

Footers – A section of concrete in contact with the ground whose purpose is to distribute structural loads from above into the ground. Footers may be deepened sections of a concrete slab or may stand alone in such places as under columns or fireplaces. Footers may also be used with or without reinforcing steel, depending on the structural loads they are supporting. (22, 33)

Freezer Circuit – see dedicated circuit (19, 33)

Gable – A type of roof common in home designs. See drawing. (49, 50)

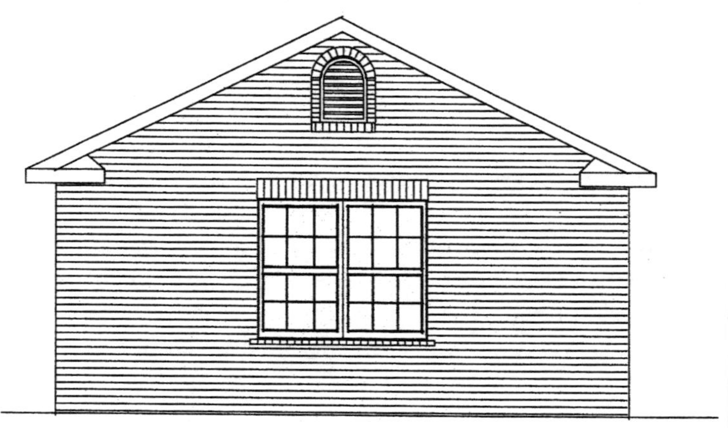

GABLE ROOF

General Liability Insurance – Insurance designed to protect the contractor (and ultimately the owner of the job) from liability that may arise from many sources such as: children injured on the job after working hours, injury to visitors on the job, etc. (61)

Glue-Lam Beam – A structural wood beam composed of smaller wood pieces glued together under controlled conditions. Because the pieces are specially selected, these beams generally have a higher strength than conventional wood beams and can be made to virtually any size. (56)

Grade – The level of the top of the ground. Existing grade is the ground level before any work is done on a building. Finish grade is the level after the building is completed. These grades may or may not be the same, depending on the homesite. (155)

Gutters – Troughs placed around the eaves of a structure to capture the rainwater which runs off the roof. This water is channeled through these troughs to downspouts, or leaders, which are pipes that carry the water to the ground. (57)

Hip - A type of roof common in home designs. See drawing. (49, 50)

Homeowners' Association – An association usually incorporated as a not-for-profit corporation whose main function is to maintain the subdivision and enforce deed restrictions. This association is usually supported by an annual homeowners' fee which is levied against each lot. (28)

Homesite – The land upon which your home will be built. This land can be either inside an engineered subdivision or on a parcel of land away from the city. (121 - 161)

Impact Fees – Fees levied by a municipality that are supposedly meant to offset the additional costs associated with new homes being built in a community. Usually these fees unfairly burden the purchaser of a new home with unnecessary and excessive costs that are not placed upon the purchaser of an existing home. (61, 66)

Infill Lot – An undeveloped parcel of land located in an area where substantial land development has taken place. (143)

Inlet – Also known as a stormwater inlet, it is the grate in a stormwater system that collects stormwater at periodic intervals on a curb and conveys the water to underground drainage piping.

Insulation – Material applied to a ceiling, wall, or floor which retards the flow of heat or noise between rooms or the exterior of the structure. (57, 62)

Ironing Board Circuit – see dedicated circuit. (19, 33)

Jurisdictional Lands – see wetlands. (152, 153)

Knock-Down Ceiling – A type of drywall ceiling which replicates antique or "hit-and-skip" plaster finishing. (62)

Lump-Sum Contract – Contract whereby the owner pays to the contractor a lump sum for all work done on the home. The price includes all material, labor, general and job overhead items such as insurance, and the contractor's profit.

Mechanical Permits – Permits required by a local agency having jurisdiction for the mechanical components of a building including: plumbing, heating and air conditioning, and electrical. (61)

Monolithic Foundation – A concrete slab, the edges of which are formed, thickened, and reinforced to serve as footings. Thus with a monolithic foundation, the foundation and floor slab are poured at the same time. (61)

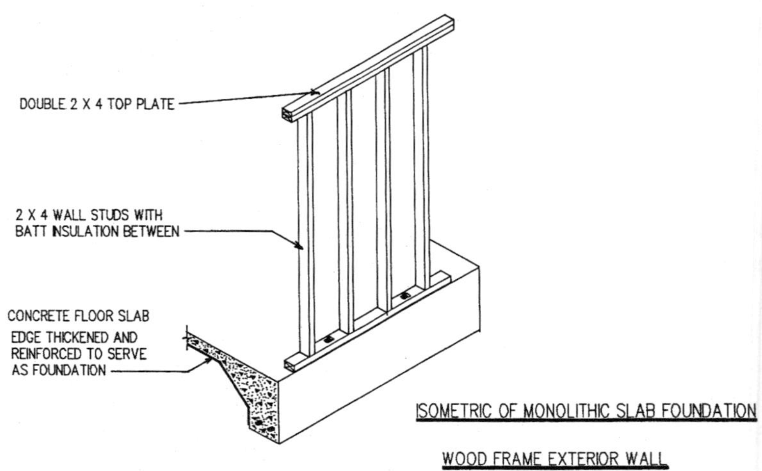

Muntins – Originally wood dividers between small pieces of glass in windows, today they are usually decorative in nature and placed between two pieces of glass in double glazed windows to simulate smaller panes of glass in traditional windows. (62, 66)

Off-grade floor – A floor which is built above the ground, usually out of wood with a crawl space underneath.

Off-site – A homesite which is located outside of an engineered subdivision. Off-site property usually does not have access to city water or sewer, which means that a well and septic tank are usually required. This coupled with extra drainage engineering and design will usually make building on an off-site lot more expensive than building in an engineered subdivision. (147 - 161)

Oriented Strand Board (OSB) – A structural wood panel made by gluing selected wood chips together with the grain oriented in certain directions for greater strength. OSB is commonly used as sheathing or decking in the same manner as plywood. (56)

Permanent Loan – The loan taken out by the owner to pay the contractor for building his home. Also known as a purchase money mortgage, it can be repaid to the mortgage company in any number of ways. (80)

Plan Approval – A time where the builder, the owners, the site agent, and the Realtor®, if involved, will review every aspect of your new home plans and your contract to ensure that the home you want is the home which will be built. (221 - 226)

Pressure Treated Lumber – Lumber that has been treated with chemicals under high pressure. This treatment makes the wood resistant to termites and decay from the elements. Due to its resistance to moisture, it is commonly used in contact with masonry or concrete work. (56)

Professional Engineer (PE) – see engineer. (151)

Progress Payments – see draw. (212)

Property Line – The boundary between two parcels of land, generally specified by length and compass bearing. (132, 133)

Punch List – A list of items on a building that need correcting. This list is made by the superintendent who inspects your home and given to those workers who need to return to fix items on the list pertaining to them. (258, 259)

Rebar – Steel bars placed in concrete for reinforcement. (55, 61)

Record Plat – The definitive map of a subdivision that is recorded in the public records of the county. All of the data necessary to locate every property line or corner in every lot in a subdivision is located on the record plat, including: right-of-ways, easements, road centerlines, and permanent control points, just to name a few. (132, 133)

Right-of-Way – A plot of land reserved by a person or group (usually government) to have the right of ingress and egress through this area. This usually means the area reserved by the county for public roads through the community. (132, 133)

Riser – In stair work, the vertical component of each step on a flight of stairs. See tread for a cross-sectional drawing of a stairway.

Roof Vents – Usually made of metal, these vents are located on top of the shingles at the ridge of a roof and ventilate the attic space of a home. (57, 62)

Rowlock – A type of brick work commonly used for sills and the tops of free standing walls. See drawing.

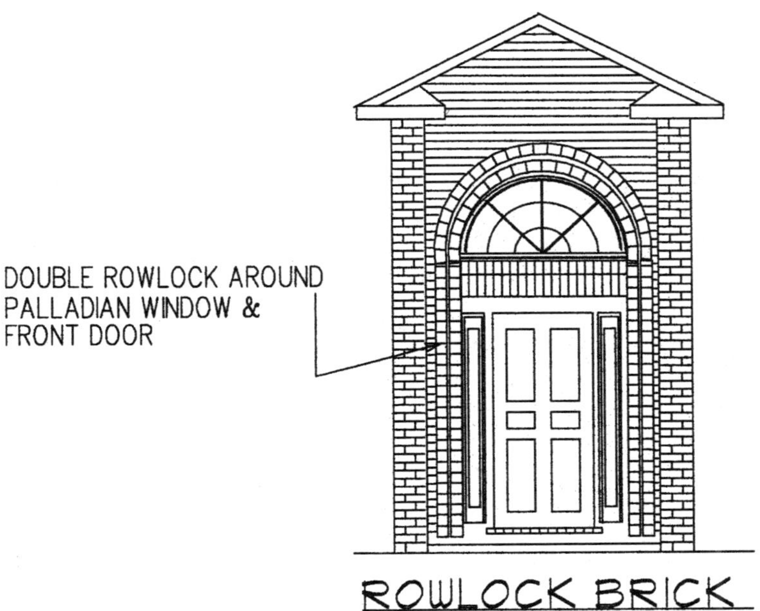

Sales Commissions – The fee paid to a real estate sales agent for bringing a buyer and seller together, obtaining a sales contract on a piece of real estate, and assisting with the completion of the sale through closing. Commission is usually paid at the closing and is usually a percentage of the sales price. (60)

Sales Plat – A sketch of a subdivision based on the record plat, showing the layout and relative sizes of homesites in the community. It does not contain the data necessary to find property corners and may not show all of the details necessary to evaluate the homesites thoroughly. (127 - 129)

Setback – The required distance from a property line to a structure. This distance is required by deed restrictions or local zoning laws. (132, 143)

Sewer Tap – Where the home's effluent pipe is connected into the sewer system. (61)

Shear Walls – Walls that run parallel to the wind and brace the walls that are perpendicular to the wind so they are not blown down in inclement weather during periods of heavy wind. (169, 170)

Simulated Acoustic (Popcorn) Ceilings – A type of drywall ceiling in which a spray application of a heavy textured paint is applied to the drywall.

Skylite – A penetration through a roof which allows light to enter, properly flashed and covered with a transparent material to prevent air and water infiltration. (57, 62)

Slab-on-grade – A home with a concrete slab floor and no crawl space underneath. The slab is elevated above the surrounding ground by placing fill dirt beneath it. (61)

Stem-Wall Foundation – A foundation system whereby a continuous concrete footing is poured and a foundation wall, usually reinforced concrete or concrete block, is installed from the footing to the floor slab. The floor slab is then poured at the top of the elevation of the foundation wall. Thus, the stem-wall foundation is installed in three steps: footing, foundation wall, and finally floor slab. (61)

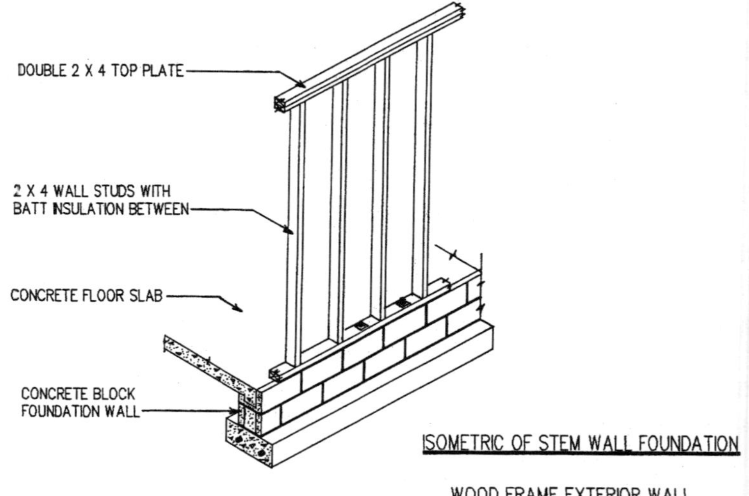

Subcontractor – An individual or a company who performs a specific portion of the work on your new home such as plumbing, heating and air conditioning, roofing, etc. He is usually paid a lump sum for this work and guarantees his work to your contractor, who then guarantees this work to you, the owner. (24, 25)

Subdivision – A plot of land which is subdivided into smaller units or homesites (lots) where homes are then constructed. In a modern engineered subdivision, items such as paved roads, city water and sewer, and cable TV wiring is included along with proper drainage engineering. (121 - 144)

Subfloor – The structural surface beneath a finished floor. A subfloor may be a concrete slab-on-grade or consist of materials such as plywood, etc. for off-grade flooring systems. (57, 62)

Subject Property – During an appraisal, the subject property is the home for which the appraiser is trying to establish a value. The subject property is compared with other similar homes in the area to establish the appraised value of the home. (20, 21)

Subsoil – The soil beneath a building site that will support the loads being placed upon it by the building. (61)

Substantial Completion – The point in the construction process where the home is complete enough so that your family could live in the home comfortably if no more work were done on the home. (258 - 261)

Summer Energy Efficiency Rating (SEER) – A rating used to compare the efficiency of air-conditioning units. The higher the SEER, the more efficient the unit. (64, 66)

Swale – A wide, shallow ditch used to convey stormwater in a subdivision. (133)

Traffic Patterns – The most heavily traveled paths people take when they walk through your new home. These patterns will affect furniture placement and the convenience of moving around each room and going from one room to another in your new home. (183, 190)

Tray Ceiling – A specialized architectural detail where the ceiling of a room is indented to make the ceiling seem higher and the room feel larger.

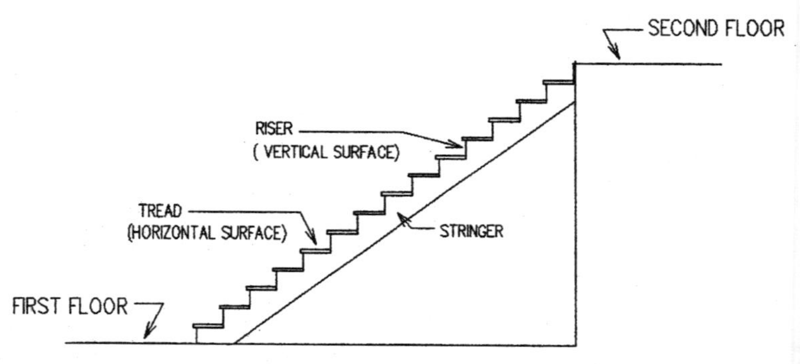

SECTION THROUGH SIMPLE STAIR

Treads – In stair work, the horizontal component of each step in a flight of stairs. The correct ratio of riser height to tread width is critical to good stair design and is governed by the building code.

Utility Hookup Fees – Also called tap fees, these are fees charged by local utility companies to tie your home into their water and sewer system. Fees are usually paid by the developer in an engineered subdivision, but may be passed on to the lot purchaser in addition to the sales price of the lot. (57, 62)

Walk-Thru – A final inspection of the home by the owner and the builder to make sure everything is complete and list items that still need attention. (251 - 265)

Wall Ties – Small strips of metal commonly used to fasten wythes or sections of brick to the backup wall material in brick veneer work. (55, 61)

Water Tap – Where the home is connected into the underground water supply line. (57, 62)

Welded Wire Fabric (WWF) – Wire mesh placed in concrete slabs for reinforcement. (55, 61)

Wetlands – Also known as jurisdictional lands, these are land areas which are environmentally sensitive and therefore are restricted by law from being developed. Wetlands are defined by the vegetation growing on the land and are usually delineated by a biologist working in conjunction with an engineer. (152, 153)

Workman's Comp – Insurance designed to compensate an employee who is injured on the job. (61)

Wythe - A single vertical section of brick or concrete block work one brick or block wide.

General Notes

<u>Order Form</u>

Fax Orders: *(904) 264-3328*

Telephone Orders: *Call Toll Free: 1-800-291-2884*

On-line Orders: *jcogd1745@aol.com*

Postal Orders: Keystone Publishing
Kelly Bowman
108 Industrial Loop North
Orange Park, FL 32073
Telephone: (904) 215-2993

Please send the following books:
(PLEASE INDICATE BOOK TITLE AND QUANTITY BEING ORDERED.)

____

____

Company Name: ____

Name: ____

Address: ____

City: _____________________ State: ___ Zip: _______________

Telephone: (______) _______________________________________

Sales Tax:
Please add 7% for books shipped to Florida addresses.

Shipping:
$4.00 for the first book and $3.00 for each additional book.

Payment:
❏   Check

❏   Credit Card (Most major credit cards accepted)

Credit Card Type: ____

Card Number: ____

Name on Card: ____

Expiration Date: ________/________

*If you have questions beyond what has been covered by this book please contact us at:
    Keystone Publishing • Dept. of Public Relations
    108 Industrial Loop N. • Orange Park, FL 32073
    or email us at: *jcogd1745@aol.com*
We will attempt to answer all inquiries within 30 days of receipt.

<u>Order Form</u>

Fax Orders: *(904) 264-3328*

Telephone Orders: *Call Toll Free: 1-800-291-2884*

On-line Orders: *jcogd1745@aol.com*

Postal Orders: Keystone Publishing
Kelly Bowman
108 Industrial Loop North
Orange Park, FL 32073
Telephone: (904) 215-2993

Please send the following books:
(PLEASE INDICATE BOOK TITLE AND QUANTITY BEING ORDERED.)

____

____

Company Name: _______________________________________

Name: ____

Address: ____

City: ______________________ State: ___ Zip: ___________

Telephone: (______) ___________________________________

Sales Tax:
Please add 7% for books shipped to Florida addresses.

Shipping:
$4.00 for the first book and $3.00 for each additional book.

Payment:
❑   Check

❑   Credit Card (Most major credit cards accepted)

Credit Card Type: ____________________________________

Card Number: ____

Name on Card: _______________________________________

Expiration Date: _________/_________

*If you have questions beyond what has been covered by this book please contact us at:
    Keystone Publishing • Dept. of Public Relations
    108 Industrial Loop N. • Orange Park, FL 32073
    or email us at: *jcogd1745@aol.com*
We will attempt to answer all inquiries within 30 days of receipt.